Deep Challenge!

The true epic story
of our quest
for energy beneath the sea

Gulf Publishing Company
Houston, Texas

Deep Challenge!

The true epic story
of our quest
for energy beneath the sea

A Trade Book
by

Clyde W. Burleson

Author of
The Jennifer Project

Deep Challenge!
The True Epic Story of Our Quest for Energy Beneath the Sea

Gulf Publishing Company
Book Division
P.O. Box 2608 ☐ Houston, Texas 77252-2608

10   9   8   7   6   5   4   3   2   1

Printed on Acid-Free Paper.

ISBN 0-88415-219-7

Photographs courtesy of Global Marine Inc.

Table of Contents

Foreword

From the bone-chilling cold of the far north to the brain-frying blast of equatorial sun, those who work offshore to find oil and gas must contend with the sea. They probe beneath the waters, always at deeper and deeper depths. They turn their bits into the antediluvian rock on the ocean bottom despite storms, hurricanes, wild waves, surging tides, and the isolation which comes from living far from home on a rig surrounded by endless water.

Hundreds of companies and thousands of people have contributed their energies, love, labor, curses, and devotion to the offshore oil industry. Many have made fortunes, some have lost millions, and still others have earned a good living by dint of hard work in difficult places. For most, there is the unending excitement of the next challenge and the satisfaction that comes from using their unique skills to win.

This book is an inside look at the marine drilling industry. It is not, in any scholarly sense, a history. Instead, the goal is to offer the flavor of a fascinating business. To accomplish this, a sincere effort has been made to follow the growth of a single company from its inception to ultimate leadership in its field. The rise of that company, its economic fall along with the rest of the oil patch, and its recovery are seen against the backdrop of the times in which this marvelous true-life adventure occurred.

Those who work in offshore oil have formed close relationships with others in the same occupation. Most can recall individuals and places from decades past with casual ease. From top executives on down, statements like, "Right! I worked with Bob on that Pan Am project off Trinidad back in '63," are heard wherever these special people congregate.

This remarkable ability creates a unique problem. Mentioning one name and not another in this text is certain to cause injured feelings, cries of consternation, and accusations of weak or poor research. In truth, a book this size could be filled with names printed in column after column on page after page and still not credit all who have given so much to this industry.

Years could be spent interviewing just the individuals who have invented or obtained a patent on some piece of equipment. And 10 volumes could not contain the number of stories—some funny, some scatological, some sad—that would come from those interviews. Preserving that lore is a worthwhile project in and of itself, best undertaken by a university or library.

This book has two goals. One is to use a look back in time, and a glance at what lies ahead, to give readers a better understanding of the important role played by the offshore oil industry. The second is to salute the efforts of those who have gone before, those who are here now, and those who will come in the future to face the deep challenge.

Deep Challenge!

The true epic story
of our quest
for energy beneath the sea

Chapter 1

Building Character: Hurricane Camille

August 16, 1969
1604 Hours — West Cameron Block 28

The night toolpusher came awake, wide-eyed and gasping for breath.

"I just lay there in the dark," he said later. "Something was sure enough bad wrong. So I tried to listen for what was the matter."

Above the incessant rumble of huge diesel engines that camouflaged all other sounds, he could pick out the clank and jangle of tongs against steel. The day-tour roughnecks were breaking out a stand of drill pipe. Then came a whir from the drawworks as the day driller pulled the 90 feet of pipe out of the hole.

Carefully, the night toolpusher's fingers felt the steel bulkhead. He noted a slight vibration. Nothing unusual. It was as if the constant machine noise had somehow bonded into the metal and made it hum.

He had worked on land rigs, submersible barges, and jackup drilling platforms like this one for more than 20 years. He'd been promoted to night pusher some months ago and had been calling the shots for his 6 P.M. to 6 A.M. crew ever since. Every sound on a rig was instinctively familiar. Nothing seemed out of the ordinary. Then what had jerked him from a dead sleep? What was sitting on his chest like a hundred-pound sack of drilling mud, making it hard to breathe?

He pressed a button on his watch, and it glowed. Five minutes after four. He had almost two hours before leading his crew onto the night tour. Time enough for a shower before breakfast. The watermaker, a desalination system that ran off heat from the diesels, had been down at the end of his last tour. So he had been unable to do more than a quick wash.

Bathing could wait. He needed to know what had snatched him awake. Switching on the light, he checked his cabin. One metal desk, one chair, one narrow bed, one clothes locker, all painted the same gray. Steel walls, floor, and ceiling, also gray. Twelve hours a day, two weeks at a time, this iron cubicle was home. In deference to his position as a night toolpusher, he berthed alone. The rest of his crew doubled up.

Had he been dreaming? Not that he could recall. Dressing quickly in a fresh, one-piece, tan coverall, he grabbed his battered aluminum hardhat from the desk.

The cabin door had no lock, so he let it swing shut behind him. Striding down the well-lit corridor, he moved fast and easy for a man his size. At the stairs, he went up, taking two at a time, steel-toed safety boots clanging on the perforated metal treads. His urgent internal alarm had receded, leaving a sense of foreboding. What was happening? He had to know.

Stepping through the iron door brought him from air-conditioned cool into a blazing August afternoon. It was more than just hot. The rig was sited 80 miles southeast of Pass Christian, Mississippi, in the shadeless Gulf of Mexico. A relentless sun hammered the handrails and deck, sending heat shimmers through the humid air.

Squinting into the harsh light, he lifted his hat and wiped an already sweaty brow, brushing gray-speckled black hair out of his eyes.

A quick glance at the rig floor confirmed what he'd heard while lying in his bunk. The day crew was tripping out of the hole in one hell of a hurry! That soft-formation rock bit they were using could not have dulled yet, so something unexpected must have happened. He watched the three perspiring men work with choreographed efficiency as they broke out another stand of pipe.

2

Scanning the horizon, he checked the movement of swells and noted the water had taken on an odd gray color. Maybe, he thought later, he had unconsciously sensed some drop in atmospheric pressure. Then too there was a faint odor that rose above the stench of diesel exhaust. It was a salty sea smell that often marked a weather change. To a Cajun raised on boats along the Gulf Coast, those signs warned of a brewing storm. He did not see any towering clouds yet. They would come, though, probably before dawn. From all indications, it would be a big blow.

August was the heart of hurricane season in the Gulf of Mexico. And they had not had a real killer since September of '65. That one took the lives of 74 people. Every drilling rig and production platform caught in its fury had been damaged.

Gulf Coasters from Key West to Veracruz accepted hurricanes as part of life. And four years was a long time between serious storms.

Wondering if this one would be bad enough to make them evacuate the platform, he headed for the day toolpusher's office to see if his boss had any special orders.

— ◆ —

August 16, 1969
1622 Hours — About 106 miles SSE of Venice, Louisiana

Driven by shrieking wind, sheets of rain hammered the high wings and fuselage of the WC-130E Hercules, striking with the power of repeated shotgun blasts. Shuddering under the onslaught, the massive plane held course and altitude, forcing its way deeper into the hurricane.

On the flight deck, the pilot pulled his shoulder harness and lap belt tighter, then checked his copilot, navigator, and engineer. The ride was going to get a lot rougher.

Behind him, the aerial reconnaissance weather officer, ARWO for short, operated his recording equipment. In a sense, the ARWO was top man on the mission because all flight crew

3

efforts were to deliver his team and their gear into the very eye of the storm.

Throttled back because of severe turbulence, the rugged old four-engine, turboprop transport held 10,000 feet. The Herk bumped along at 250 knots, considerably below optimum cruise speed.

Each 30 seconds, automated sensing gear measured outside air temperature, dew point, aircraft altitude, and barometric pressure. In addition, the ARWO took notes on icing, visibility, cloud types, cloud cover, and ocean surface winds. Each 400 miles, and on every pass they would make through the storm, the dropsonde system operator, or DSO, would release an instrument cluster. The dropsonde, a cylinder less than a foot and a half long and about the diameter of a baseball, was deployed with a parachute to slow its fall. Descending toward the frenzied Gulf waters, the package radioed the DSO a detailed vertical atmospheric profile.

On this reconnaissance mission, the Hurricane Hunters' aircraft made only two trips into the storm's center. Pressure was 901 millibars, and surface winds exceeded 175 knots—more than 200 miles per hour! Mechanical trouble cut their second intrusion short.

About 250 miles south of Mobile, Alabama, just entering the eye, the number three engine failed. With the needle indicating internal turbine temperature off the scale, the crew had to shut it down before it tore itself apart. Which would probably have taken the wing along with it. Losing the engine at that critical moment did little for their confidence.

The flight out was hair raising. Pressure from the wind was intense enough to make the plane's metal skin rub against the airframe, filling the Hercules with a noise like moaning. Flying on only three engines forced them to a lower air speed, which proved to be a help. The storm radar, however, which usually could be relied on to plot the least turbulent flight path, was almost useless. The entire screen was a splotchy red, indicating chaotic weather in all directions.

After 18 minutes of harrowing flight, they broke through and found themselves in relatively clean air. Hurricane-force winds extended out a full 60 miles. The crew's sighs of relief had barely subsided when the ARWO made a sobering announcement. They had just flown a crippled plane through Camille, the second most powerful hurricane to threaten the U.S. mainland during the 20th century.

—◦—

August 16, 1969
2000 Hours — En route to Venice, Louisiana

Captain Loyd Dill, skipper of the *Glomar II*, the first vessel to be designed from the hull up as a drillship, checked the clock and chart on the table in front of him. His crew had been working in South Pass, Block 89, about 50 miles southeast of Grand Isle, Louisiana, engaged in a drilling program for Humble Oil & Refining Company. As the weather worsened that afternoon, they received reports on Camille's position and estimated course. The decision to run for safer, more protected waters of the Mississippi River had been made without hesitation.

All hands eagerly fell to. Time was short and they wanted to be underway. One crew brought the bit and drill string out of the hole and then assisted with retrieving the marine riser pipe. Helped by the supply and anchor-handling boat, the *Rita Candies*, and the tug, the *Frank Candies*, another crew detached the wire mooring lines that had been holding them on station during the drilling operation. By 8 P.M. they were heading for the southwest pass, which would allow them entry into the river.

Early next morning, they neared the small town of Boothville in Plaquemines Parish. One look at the dark, greenish sky told more than the latest weather bulletins. Time was running short. Camille was on her way.

With the *Rita Candies'* support, Captain Dill watched as two 20,000-pound anchors, one at the bow and one at the stern, were set into the bottom. The location had been selected because they could also use nylon hawsers to moor their ship alongside

5

steel pilings near the river's west bank. As they were finishing the operation, gale-force winds whipped the usually placid waters with a wild intensity.

Between 1:30 P.M. and 8 P.M., the hurricane grew in strength, assaulting the *Glomar II* with 160-mile-per-hour gusts. Storm tides raised the river's water level dramatically, and there was great concern the levee might give way, inundating the surrounding hamlets. Somehow the levees held, but the towns were swamped anyway. A hammering, ceaseless rain, combined with sheets of water sucked from the tide-swollen river and blown over the banks into low areas on the west side, produced devastating floods.

The men on board the *Glomar II*, who had raced from the Gulf for shelter, now found themselves less than 20 miles from the hurricane's center. They had done what could be done. Now they had to rely on their stout ship and ride it out.

~ ~

August 17, 1969
0422 Hours — West Cameron Block 28

An intense, blasting rain pelted the night. There was no way to judge how fast the shrieking wind was tearing across the rig's exposed drilling deck. All sensors of the Easterline Angus wind-speed recorder had blown away an hour earlier. The last reading was 172 miles per hour.

The night toolpusher stared through the dank dimness at his crew seated with him in the galley. They were using emergency lighting because all generators and big diesels had been shut down. Running the engines was too dangerous. With so much water in the air, some of it could be sucked into the cylinders, causing a failure that might trigger a fire.

Backup electrical power provided limited air circulation but no air conditioning. So the room was stuffy and close with the odor of cigarette smoke and sweat. Opening a door or ventilation port was out of the question. Raging waves, breaking through the platform's massive open-truss support legs, added to the flying water. And the wind, which had smashed outside structures like the corrugated steel toolhouse on the drilling floor, turned debris into deadly missiles.

Earlier, just after nightfall, the toolpusher in charge of the rig had called all 45 men on board to a special meeting. He told them there was good news and bad news. The good news was that the hurricane's center was now expected to pass 50 miles to their east. For the bad news, he read a radio message stating storm-induced fog had prevented their evacuation helicopters from taking off. Every man on the platform understood what that meant. Even at dusk the seas had been too fierce for crew boats or larger supply vessels to make a pickup. So they had no choice but to stay on the rig. If the violent winds threatened to destroy that shelter, their only hope was to take a chance inside the emergency escape capsules. Those watertight, fully enclosed, clamlike fiberglass lifeboats were supposed to be unsinkable. But they knew that other storms had washed men overboard before they could reach the capsules. Worse, people had drowned in water trapped inside wave-tossed lifeboats. No one knew for certain how the boats would fare in seas whipped by a Category Five hurricane. No one had any desire to take part in that test.

As hours passed, winds began gusting with greater force. Revised position plotting, using information from one of the Hurricane Hunter aircraft, wiped out the earlier good news. They were smack in the path, and Camille was approaching fast. By 0930 hours, when it should have been daylight, the world was pitch black.

Trying to maintain morale, the catering staff began handing out sandwiches and pouring coffee. All activity stopped when the walls of the rig's enclosed living quarters began to flex. Welded sheets of steel started bowing in and out like the sides of a live animal panting for breath. A horrible shriek of tearing metal, loud despite the storm's insane howling, announced that part of the superstructure had been ripped away. Water now streamed down the inside walls of the galley. As damage mounted, the idea of using the emergency escape pods appeared better and better.

Finally, after what seemed a terrifying eternity, wailing winds abated. An eerie stillness descended as the day brightened.

One man went to the hatch and listened. Hearing nothing, he cracked it open and peeked outside. He saw patches of blue sky and rays of sunlight. To the north and south, and all around, was an ugly wall of black clouds, towering from the water up to heaven. The rig was smack in the eye of the storm.

Taking advantage of the momentary calm, the men ran a quick platform inspection. Damage was horrendous. The crown block, high in the derrick, had collapsed on the rig floor. Drilling line was splayed everywhere in tangles around the drawworks. The broken derrick was twisted beyond repair, and quarter-inch steel sheeting had buckled. A three-foot gash, slashed through one bulkhead, still held the wind-driven slab of metal from the doghouse that caused the destruction.

Their real problem, though, came as a shock. Instead of being level, the whole platform now slanted several degrees. Which meant one of the support legs might have bent. Or more likely, the spud can on the end of a leg had dug itself deeper into the seafloor. Either possibility spelled serious trouble because the tremendous weight of the entire platform was not evenly distributed. The leg that was burrowing or flexing was being forced to take a greater and greater load. Under pressure from the wind and waves again, it might buckle. If that happened, the rig could topple. Everyone on board would die.

The men were sweating, even though it wasn't really hot. A light puff of breeze almost caused the group to shiver. The hurricane was heading northward and they were about to get the rear wall, which meant winds from the opposite direction. That change could put different stresses on the platform.

Back in the galley, the toolpusher explained the situation. And their options. They could stay—or go for the escape capsule before conditions deteriorated again.

For several minutes, everyone was deep in thought. Finally, the roustabout foreman, a crane operator, spoke up. He was going to remain on the rig. If anyone wanted to try the sea, he would lower their capsule after they had boarded and locked the hatches. The roustabouts and roughnecks he bossed allowed as how if he weren't going, they would stick with him.

Not much discussion ensued. Each man was holding his private fears. The thought of being tossed around in storm-churned seas was horrifying. In comparison, the jackup platform was an island of stability. The slanting deck, though, offered its own terrors. One after the other, they silently made up their minds. The best bet was to trust the tons of steel lattice that held them above the angry seas.

Once that decision was made, there was nothing to do but wait—and watch the sky darken again as winds reintensified.

Each second seemed to last a minute. Each minute, an hour. Each hour, a lifetime. When the jackup began creaking and groaning as its metal was burdened beyond design limits, hearts stopped beating and men held their breath. One giant gust sent a sickening quiver through the massive structure.

The fierce wind reached maddening velocities and stayed there. Anything that could be blown away was gone. So there were no crashes or bangs from debris being slammed into the bulkhead. In place of that came sustained, metal-ripping screeches that dug at every man with the impact of fingernails dragging over his bare brain.

Then one of the crew looked at the cold coffee in his cup sitting on the table in front of him and almost choked. It wasn't level. The coffee was higher on the right than the left. The platform had tilted some more. That little discovery was a real character builder.

Across the hot, sweaty room, another man came to the same conclusion and raised his voice, "Hail, Mary, full of grace...."

Many others joined him, and when they were done, a different man began a hushed "Our Father, who art...." Brave men, hammered by nature's wildness, prayed together, for themselves and for the safety of their families.

—•—

August 17, 1969
1530 Hours — West Cameron Block 28

Several anxious, sleepless hours later, as the sky lightened, wind velocity fell. The driving rain became a downpour, then a shower, then ceased altogether, leaving rivulets and puddles on parts of the jackup's iron deck. They had passed through the winds of hell and survived.

Since the storm was between them and the coast, they knew it would take time before anyone could come to their assistance. By rerigging their radio antennas, they were able to contact stations in Brownsville, Texas, and Tampa, Florida. Assured that every effort would be made to notify their loved ones, they started the diesel-powered electric generators and set out to restore order.

It would be a long while before this rig could be used for drilling again. The derrick would have to be replaced. And salt water had already begun its corrosive work on electrical connections as well as machinery. Worse, all three support legs had actually been moved by wind and wave action. Shifting position had snapped the conductor pipe to the seafloor and damaged the blowout preventer stack slung below the platform.

August 18, 1969
0700 Hours — Venice, Louisiana

On board the *Glomar II*, most of the men had been up all night. With first light, some sense of the devastation around them became apparent. As the day progressed, Captain Dill and Bob Bright, Humble Oil's company man, went out to scout the stricken area.

The town of Venice was under water. It had been lashed by Hurricane Betsy a few years earlier and rebuilt. Now this. The new $3.5 million Getty terminal and gasoline plant were wiped out.

On the *Glomar II*, damage was slight. The drawworks shelter, made of corrugated steel, was gone, as were several radio antennas and a mooring bitt. After quick repairs, the ship headed back to her drilling location in South Pass Block 89. Marker buoys were still visible. In 36 hours they were remoored and ready to return to work. Members of the crew joked that if you've seen one hurricane, you've seen them all.

You hope.

All told, Camille knocked out an estimated 300,000 barrels of production a day and cost the oil industry $100 million in damage to equipment. At least three large platforms were swept away, and more than 15 rigs of various types were partially destroyed.

After landfall, Camille continued to ravish parts of Tennessee, Kentucky, West Virginia, and Virginia before slipping off into the Atlantic Ocean. In all, 258 people were killed and 68 were classified as missing. Reports in oil industry magazines from that time do not indicate any deaths of those working offshore.

Even so, drilling for oil and gas under hundreds or even thousands of feet of water remains a risky business. Risky for those companies that bet millions upon millions of dollars on a single well. And risky for the men who devote a large part of their lives to making that well possible.

How can searching for oil and gas justify such an enormous gamble?

While the stakes may be high, the money is right. Wages for those who work on the rigs are exceptional as are returns on capital for investors willing to take the risks.

To view money, though, as the sole motivator is misleading. Offshore professionals at every level, from roughnecks to CEOs, share an enthusiasm for their work. There is an exciting mystique about searching out, then producing seabed petroleum deposits. And a big part of the allure comes from understanding that oil and gas, developed in commercial quantities, has changed the course of human history.

To test the accuracy of that statement, it is necessary to go back to 1859. That was the year a single, simple discovery altered the entire structure of our world.

Chapter 2

In the Beginning:
From Titusville to the Deep Blue

As surprising as it may seem, there are valid grounds for dividing human history into two distinct periods. The first contains all that happened up to the year 1859. The second encompasses what has occurred since.

During the ages prior to 1859, our ancestors learned to chip flint rocks into useful tools. They also developed the ability to control fire, cultivate crops, domesticate animals, and smelt metals. Each of these discoveries played a vital role in the development of our civilization. Each altered the very soul of society.

The breakthrough made on a hot August day in 1859 was equally momentous. At the time, few took notice. Even the newspapers in Crawford and Venango Counties, located in Pennsylvania, paid scant attention. Yet this one accomplishment was to redefine the material standards of human life. And we are still adjusting to that transformation.

In 1859 we learned how to drill for petroleum. And a plentiful supply of petroleum brought light to the world, then ended humanity's age-old reliance on muscle, wind, coal, and water power to create the necessities of life.

Petroleum, not just for use as a fuel but for lubrication as well as many products, allowed the development of cars, trucks, planes, trains, rockets, telephones, computers, and pantyhose. Petroleum made those and other inventions, from the generation of electricity to tractors for working our farmlands on a never-before-imagined scale, practical realities. Petroleum gave us the means to build and operate machines that have changed the earth's face. Petroleum products are at the heart of all modern technology. Petroleum, allied with human imagination, has given us the ability to leave our planet's confines on voyages into near and outer space. Petroleum has allowed us literally to reach for the stars.

The individuals who sank that first well had no notion of starting a petroleum revolution. They acted, precisely like those who would come after them and drill oil wells by the thousands, from ambition. Although they did not foresee that the "rock oil" they sought would reshape society, they hoped to sell a whole lot of it. Which was why they drilled a well in the first place.

To fully appreciate what happened, it is necessary to look backward in time.

America of the mid-1800s was very different from the America of today. In fact, well into the first decade of the 20th century, people lived much the same as their grandfather's grandfather's grandfather. In what some refer to as "the good old days," only one home in seven had a bathtub. A hand-operated water pump inside the kitchen was progress. Daily sanitary needs were accommodated by chamber pots or the little house out behind the big house. In cold weather, heat was supplied by wood or coal-burning stoves and fireplaces. There was no air conditioning. When it was hot, the people were hot, without even fans to stir the torpid air.

Electric lights were science fiction. Homes and businesses were illuminated, as they always had been, by candles or lamps burning oil rendered from animals or vegetables. The light was dim and flickered.

Muscle power performed most work. Muscle—horse, ox, donkey, mule, or human—plowed the land, transported goods from place to place, and took people to church on Sunday.

A few brave ships ran on steam engines, but fewer still were intrepid enough to dispense with sails entirely. Steam also drove the railroads and had just begun to turn the wheels of industry. All applications of steam power suffered from the same problem. Parts that rotated or moved back and forth rubbed on one another and required lubrication for reliability and extended life. The only lubricants available were either processed from plants, like castor oil, or rendered from animal fat. None was well suited for the purpose; all broke down readily under exposure to high temperatures, and even the best required constant replacement. Frequent manual oiling of all moving parts and bearings produced only limited protection, resulting in reduced performance.

On railroad cars, where rolling wheels could not be reached for oiling by hand, bearings were packed with animal grease and hope. Failed lubrication resulted in a "hot box," as the condition was called. It was not uncommon for a wheel to seize solid on its shaft, and with sparks flying, be drug along the track until it glowed red, then broke. Busted wheels often led to derailments.

Engineers and chemists worked diligently on improving the lubricating qualities of available oils and greases. Only restricted progress was made because of the limiting nature of the basic raw materials.

The United States, and the world, had a serious need for a better lubricant. And since whales were being hunted almost to extinction, the progress of society also demanded a different oil to light residences, shops, offices, and factories. On those needs hinged the fate of the industrial revolution—and the capability to lift mankind beyond the dependence on animal muscle.

Few imagined that the answer lay in petroleum. In the many places where oil seeped from the earth, it was considered a smelly, worthless bother.

In time, petroleum proved to be an awesome solution. By the early 1900s, America would run on it. The rest of the world would soon follow. Exploring for petroleum, drilling for petroleum, producing petroleum, refining petroleum, and selling petroleum products were activities destined to form one of the biggest industries on earth.

The business actually had a strange beginning. In 1849 a druggist, S.M. Kier, was partner with his father in a saltworks supported by brine wells near Tarentum, a town several miles north of Pittsburgh, Pennsylvania. In addition to brine, the wells threw out small quantities of petroleum, which were unceremoniously dumped into a nearby canal. Somehow, the petroleum accidently ignited, and the resulting conflagration threatened both life and property. Extinguishing the flames, which burned on top of the water, proved to be an ordeal. When the smoke cleared, Kier, not wanting to repeat that experience, instigated a new disposal program. He ordered the unwanted oil to be discarded on bare ground.

Frugal by nature, Kier was bothered by this waste. So he came up with the notion of bottling the sharp-smelling stuff and selling it as a cure-all with "wonderful medical virtues." A flyer for the product stated that the oil had come from "400 feet below the earth's surface." Door-to-door vending by agents moved more than 600 bottles a day, which was a tribute to the men's sales ability. Not to mention the gullibility of their customers.

Even with that brisk trade, the expense of labels, bottling, and commissions allowed Kier little profit. So in 1852 he switched marketing tactics and began offering his elixir through drugstores. Sales fell and he was left with more petroleum than he could use. Recalling the fire, he decided it might make a fuel—if he could correct its properties. So he called upon his druggist's skills, distilled a batch, and peddled the result for $1.50 per gallon. This unique sequence of events made him the first man in America to refine petroleum commercially.

More serious work, however, was underway. A complex system for extracting crude oil from suitable coal, and then refining the result into "coal oil" and paraffin, was already a profitable venture.

Seeing the sales potential of this new product, a Canadian, Dr. Abraham Gesner, sought a simpler way to produce a similar commodity. Dr. Gesner was a man of many talents. At one time in his past, he had exported livestock to islands in the Caribbean and been shipwrecked. Having had enough of that, he migrated to London and studied medicine. Then he returned to Canada and became a geologist. In 1854 he invented and pat-

16

ented a method of distilling petroleum to produce a clear, flammable oil, ideal for use in lighting. Since every home and place of business would need several lamps, there was a considerable potential market for the new "kerosene." Dr. Gesner coined that name from the Greek word "keros" for wax because paraffin, that waxy substance, was also a by-product of his process.

Gesner helped a group of investors construct a kerosene refinery in New York, and five years after filing his patent, the plant was producing thousands of gallons every day.

Compared to whale oil, rendered from those huge denizens of the sea, both coal oil and kerosene gave a much brighter, cleaner light. And whale oil was becoming scarce as more and more whales were killed for their blubber, which was being systematically stripped and boiled to meet growing demand for artificial lighting. Needless to say, those in the whaling and whale oil refining industry did not appreciate new competition. Some attribute to this group the ugly rumor that petroleum oils were dangerously explosive. Unfortunately, there was more than a smidgen of fact in the allegations. Poorly refined petroleum contained volatile agents, including gasoline and naphtha, which could and did explode. Better quality control and improved refinery techniques eventually corrected this problem.

The real trouble lay in the fact that kerosene and coal oil, when burned in common lamps, gave off an acrid, dense black smoke. So a special lamp was required. First used in Vienna, the clear glass chimney and wide, flat wick took care of the smoke as well as the smell. Imported, improved, and priced right, the revised lamp became the foundation for a complete line of retail products called "kerosene goods." With this impetus, coal oil began to sell in quantity along the East Coast.

The scene now shifts to Dartmouth College, in Hanover, New Hampshire, a place where not a drop of oil has ever been found in the ground.

George H. Bissell, a determined man with a long face and full, scraggly black beard, had supported himself since the age of 12. He'd worked his way through Dartmouth by writing articles and teaching. After graduating, he had become a professor of

Latin and Greek, a journalist, and then superintendent of public schools in New Orleans. He spoke and read a half-dozen languages before beginning his study of law. After an illness, he decided to return north, and while passing through Pennsylvania, came across the primitive oil industry. Oil, floating on top of brine at a saltworks, was skimmed off or soaked up into rags which were then wrung out over open tubs.

Arriving in New York, he established a legal practice and on occasion went to see his mother, who lived in Hanover, New Hampshire. During one visit, he stopped at his old college. Bissell called on Professor Crosby, his friend and former counselor. While they talked, he was shown a bottle of petroleum that had been left by another Dartmouth graduate who was a country doctor. The rock oil came from a spring on property owned by a lumbering company near Titusville, in northwestern Pennsylvania, near where Bissell had stopped. Neither Bissell nor Professor Crosby was aware of Dr. Gesner's recently filed patents or the earlier petroleum adventures of Kier. Still, both men were struck by the similarity of this rock oil to coal oil.

Interested, Bissell contemplated its commercial possibilities. He recognized the medicinal uses for rock oil, which was prescribed for humans and animals alike. He also knew it was flammable. That gave him the idea of using the oil for lamps.

Needing more information, Bissell agreed to send Professor Crosby's son to Titusville on an inspection tour.

The trip to Pennsylvania and back required several weeks, and during that time Bissell managed to bring together a group of financial supporters. After receiving a favorable report from the younger Crosby, Bissell and his backers then retained Benjamin Silliman. Silliman, a professor at Yale College, was a famous scientist of the day. For a fee, he agreed to make further tests. An encouraging account from the chemist, who would do a fractional distillation of the oil, would then be used to bring in more investors.

To be ready, Bissell and his law partner, Jonathan G. Eveleth, bought 105 acres of Pennsylvania land that included the petroleum spring. Along with some colleagues, they formed the Pennsylvania Rock Oil Company to market petroleum for lighting, medicinal aids, and possibly as a lubricant for machinery.

There was only one difficulty. When it came time to pay Silliman and receive the report, Bissell and his group didn't have the money. Without the $500 plus, Silliman refused to release his study. Finally, after days of scrounging, Bissell found an associate who would put up the cash, and the document was acquired. Dated April 16, 1855, the paper was rushed to a printer. As soon as the ink was dry, Bissell had copies in the hands of potential backers.

Selling stock in the new venture was a challenge. Money was tight, the idea was unproven, and neither law partner wanted to spend much time away from a lucrative practice. In addition to the other problems, laws of New York state in that period made stockholders of a joint stock company liable for debts of the company.

Connecticut law held no such liability, so the firm was reorganized with New Haven as headquarters. Stock was eventually sold.

The goal of the company was to sell oil taken from their land to kerosene refineries. If they could get enough oil, they would be able to undercut the price of feed stocks and make a tidy profit. The trouble was in amassing sufficient oil to make their ploy work. The usual methods of collecting oil as a by-product of salt wells hardly guaranteed the required supply.

Then serendipity struck. Bissell spotted one of Kier's petroleum patent-medicine flyers in a drugstore window. A picture in the ad depicted derricks and holding tanks that were used in the saltworks. Aha! The oil they needed would come from their Pennsylvania property by drilling more wells on it.

Re-energized, the two attorneys reviewed their holdings and discovered a mistake had been made when transferring the deed. The wives of the original owners, Brewer, Watson & Co., had not signed the document, which placed a potential cloud over the title. This oversight had to be rectified. And discretion was demanded to prevent the ladies from requesting additional payment for their signatures.

Bissell discussed the difficulty with James Townsend, a New Haven banker who had helped peddle the stock. Townsend, in turn, knew a man who might be prevailed upon to go to Titusville and remedy their error.

Edwin L. Drake was an unemployed railroad conductor, recuperating from bad health, who lived in the same hotel as Townsend. A tall, rather handsome man with a pleasing manner, Drake was 38 years old. He dressed well, kept his beard trimmed in the fashion of the day, possessed a dignified charm, and told a good story. None of these traits was a disadvantage, considering his mission to win the women's cooperation.

Another strong point in Drake's favor was the train pass he held that would let him travel free from New York to Titusville and back.

To build his prestige and give their emissary added stature, Bissell and his partner conferred the title of "Colonel" on Drake, which was warmly accepted. To support the bogus accolade, letters addressed to Colonel E.L. Drake were mailed so as to arrive in Titusville ahead of the addressee.

Drake hit the mud street of the tiny lumber town, population under 150 souls, in 1857 and promptly went to work on correcting the title matter. He must have been engaging because, in just a few days, the wives appended their signatures to the deed, gratis.

During his stay in the area, Drake saw raw rock oil used as a lubricant in a sawmill. Familiar with trains and the problems caused by poor lubrication, he became interested in the substance. In a short time, he was a convert to the potential of rock oil.

He visited the brine wells at Tarentum, 80-odd miles away, and witnessed the primitive oil collection methods in use. Then he began reflecting on a major problem. The quantity of petroleum he envisioned being needed was vastly more than was flowing naturally from the salt wells. So boring more of those was not the answer. He needed an oil well.

Returning east to meet with Bissell, Eveleth, and Townsend, he discussed his oil well idea with several well-informed individuals. All regarded the notion as impractical. According to the thinking of that time, oil was a seepage from coal and did not exist alone. Despite this, Drake remained firm in his belief.

Investors in the Pennsylvania Rock Oil Company decided they would give the persistent man an opportunity to develop the production they so badly needed in order to make a profit. They formed a new corporation, Seneca Oil Company. The name came from an Indian tribe which had been associated with surface-seeping petroleum since the earliest days of exploration. Drake was named general agent, and Seneca was given a lease on the oil land owned by Pennsylvania Rock Oil Company.

In 1858, with business matters under control, Drake returned to Titusville where he tried digging as a means of increasing flow from an oil spring. Failure at this made him even more determined to bore. In spite of limited funds, he managed to acquire a steam engine, boiler, and other equipment. He was ready to commence drilling. There was only one difficulty. He had no experience in boring a hole into the earth. And his oil well project was ridiculed by many in the community, so he found little cooperation. At least two experienced drillers agreed to take on his job, then failed to appear for duty.

Winter was coming, so he picked his site with care and then directed construction of a wooden engine house topped by a fully enclosed plank derrick. The spot he selected was on an island at the convergence of Pine and Oil Creeks in the Cherrytree Township, Venango County, Pennsylvania. He also became the world's first company man, or on-site representative of those who had obtained the ground lease and put up the money for a well.

It took him until the spring of 1859 to find someone to drill. A blacksmith from Tarentum, William A. Smith, accepted the challenge. "Uncle Billy" arrived in a wagon with his two sons, who were to serve as "tool dressers," and his daughter, who would act as cook. He also brought the needed well-drilling implements, including bits and reamers he had hammered out by hand at his own forge.

In mid-April all was ready, and work began. Four months later, in August 1859, they had gained a depth of 69 feet. They had also drained their finances.

The shareholders had long since soured on the idea of

21

boring. One, the banker Townsend, remained steadfast, personally paying some of the bills. When Townsend's confidence finally faded, he sent Drake some money and a letter instructing him to stop further drilling, pay outstanding bills, and return to Connecticut.

On August 27, 1859, Drake had not received the letter. He was, nonetheless, discouraged. Then, at the 69-foot depth, their drill suddenly dropped six inches. Since it was a Saturday, they stopped, planning to resume work Monday. On Sunday, Uncle Billy was unable to stay away from the site and found the bore hole almost full of oil. Colonel Drake had his well. Others, like George Bissell, who rushed to Titusville with funds to lease or buy farms in the area, would become wealthy from the riches of petroleum.

Edwin Drake did not. Like so many pioneers before him and oilmen who would come later, he died broke.

Businessmen and speculators who caught wind of Drake's well also caught oil fever. Within a year, the landscape was covered by wooden derricks. Not all seekers, though, were as fortunate as the colonel.

Henry R. Rouse, along with fellow investors, was satisfied with the progress of his well at the Buchannan farm, east of Oil Creek. Then, during the evening hours of April 17, 1861, a rumble came from deep in the earth, and the well suddenly blew out of control. Oil spewed upward into the air, raining back onto the ground in torrents. The escaping oil and gas ignited, and the deadly explosion was heard miles away. For three days, despite best efforts, the wild well continued to flare. It was finally doused by covering it with dirt to smother the flames. By that time, 19 people had died and 10 others were seriously burned. This was the first major oil exploration disaster. It would not be the last.

After a somewhat slow start, demand for oil boomed, initially for lamps and industrial lubricants, then as a boiler fuel. By 1863, in the midst of the Civil War, more than three million barrels of oil were being taken from the ground each year. And that was only the beginning.

Thomas Edison's invention of the incandescent light bulb,

though, marked the end of kerosene demand. In 1882 the Edison Illuminating Company had a test installation in New York City. By 1902, 18 million light bulbs were being used in America. And more homes were being wired every day. The transition was slow, but inexorable.

Steamships and locomotives began converting from coal to oil because oil delivered improved performance. This change bolstered the fledgling petroleum industry. It was gasoline, however, that reignited oil fever.

Henry Ford, who once worked as an engineer for the Edison Company in Detroit, had resigned to devote his attention to designing a horseless carriage. When the year 1900 ended, there were fewer than 14,000 automobiles in America and just 144 miles of paved roads. Ford founded Ford Motor Company in 1903 and turned the automobile from a luxury to a necessity. In 1917, less than two decades later, 1,745,792 automobiles were manufactured during a single year. And the price of gasoline, which at one time was a penny or two per gallon, jumped to rival that of the then more expensive kerosene.

The motorcar created an unprecedented need for fuel. Demand was so great that the original Pennsylvania oil fields could not possibly fill the supply lines. New sources were essential because petroleum, and what could be distilled from it, was becoming the basis for a new society. The productivity of each man or woman was about to be increased tenfold.

Such unprecedented need created a serious problem. There were only a limited number of places in the U.S. where oil and gas oozed to the surface. Aside from those rare spots, no one had a clear idea where to explore. So exploitation of available sites received top priority.

Reports of "bituminous effusions" drifted east from California, which had become a state in 1850. The news triggered a rush for "black gold" that made the 1848 California rush for real gold at Sutter's sawmill seem puny in comparison. By the early 1860s, would-be J.R. Ewings had acquired mineral leases for hundreds of thousands of acres.

In 1864 promise of another fee from rock oil sent the ubiquitous Professor Benjamin Silliman west to California. His assignment was to examine potential areas of petroleum production. His report to the Pennsylvania Railroad only added to the swelling oil speculation frenzy.

H.L. Williams, a member of a spiritualist cult, completed two wells on the thousand-plus acres he held, known as Ortega Rancho. Then, in 1889, he took time off to develop a town for those who believed, as he did, in spiritualism. Seeking a name for the new settlement, he hit on "Summerland" after the title of a popular spiritualist publication. Two years later, the town had a temple, donated by Williams, along with about 60 homes. There were also nine gas wells which had been drilled to fill residents' needs and supply natural gas to the nearby city of Santa Barbara.

By 1895 Summerland had 28 wells and was booming.

When better-producing sites were found by drilling on the beach right up to the waterline, the inevitable happened. In 1897, to get nearer the motherload, Williams built a wharf and drilled the first well over the sea.

Foot by hard-won foot, oil seekers strove to drill away from land in deeper and deeper water. The first tentative steps off the rickety wooden piers produced the beginnings of a new technology called marine drilling.

The next big play on submerged land came from halfway across America. A few years before the Summerland boom, in 1870, a water well being drilled for an ice plant in Shreveport, Louisiana, produced natural gas, which was promptly harnessed for illumination. Eagerness to drill where there was any prospect of an oil discovery brought the Shreveport area under serious scrutiny.

Several miles north of the city, on the Texas-Louisiana border, deep in pine woods, lay Caddo Lake. The Caddos were a large Indian confederation of the mound-building, Mississippian culture. According to their legend, the sacred, dark body of water had been created when the Great Spirit saved the tribe from annihilation during a flood.

Caddo Lake was indeed mysterious. Sounds, like low bellowing, could be heard from the depths. And bubbles rose to the surface, then burst, leaving behind a rancid odor. Those same bubbles, if ignited, would burn fiercely, without stopping. At places along the heavily wooded shores of the long lake, black oil made small puddles.

Regardless of the eerie nature of the area, those searching for petroleum recognized the lake's potential. They were also daunted by the added cost and complexity of drilling under water. As a result, seepages on dry land received more immediate attention. The Caddo field was opened in 1904 with the sinking of one well. Drillers found oil, then abandoned the project when it would not flow.

The notion that recovering oil from the field might be difficult was quickly laid to rest. In fact, the pressure of underground natural gas almost spelled its end.

The Producers No. 2 well was the first of many to blow wild. In May 1905, Producers No. 2 blew out in a ferocious explosion that ripped a deep crater and then shot flames into the air, incinerating the derrick and related machinery.

For the next half-decade, one well after another blew out. Some burned for years with flames that lit up the darkness and could be seen 25 miles away. The unending roar of escaping natural gas was as loud as a freight train. Residents of the town of Caddo lived with that noise day and night. And every 24 hours, 70 million cubic feet of gas spewed into the sky.

For many, the worst disaster occurred on May 12, 1911. The Harrell No. 7, just completed, let loose so fiercely that sparks from sand being blasted up the well bore ignited the gas. A man died at the scene and three others were seriously burned. A column of fire spouted 75 feet into the air and, because of a clogged valve, sprayed out to all sides, making the well almost impossible to approach. One scheme to extinguish the fire called for a cannon. In addition to consuming all oxygen and possibly snuffing out the flames, an exploding cannonball could blow off the wellhead, allowing the fiery stream of gas and oil to shoot straight up. Then the burning well could be approached by firefighters.

Unfortunately, the cannoneer missed. So other measures were taken. It required more than 40 men, working in shifts, to dig a 20-yard tunnel deeper than 10 feet, then lay a pipeline to intersect the well. They did it in a week of intensive labor and were able to divert the gas flow enough to blow out the fire. The well produced a loss of nearly $200,000—a lot of money in those days.

When Star Oil Company's Loucke No. 3 well ignited, the burning oil was estimated to exceed 30,000 barrels a day. Loucke No. 3 was considered by experts of the time to be the biggest oil well fire in U.S. history.

At first glance, those early pioneers appear abominably wasteful. In their defense, they had limited technology to help extinguish an inferno fed by gas at unprecedented pressures. Plus they didn't know much about preventing a blowout before it happened. To make matters worse, experience had shown that oil often flowed after gas pressure was released. Since they were searching for oil in the first place, they had little incentive to combat the burning well sites or stop the spouting gas. The fact that there was no market for such huge quantities of natural gas also played a role in their thinking.

In spite of the laissez-faire attitude of the state of Louisiana, waste and destruction became so great that the Caddo Parish authorities had to act. Aided by support from the U.S. government, they passed some of the earliest laws regarding petroleum conservation. The new regulations did not completely resolve the Caddo field problem or totally stop the squandering of gas but did serve to improve what was quickly becoming a desperate situation.

In spite of the very real dangers, well after well was "spudded" in and sunk. By 1910 wells extended completely along Louisiana's shore of the lake. Prospects of work or dreams of being an oil baron brought thousands into the area. Lease hounds arrived baying for bargains, promoters as well as speculators talked hysterical amounts of money, and roughnecks earned almost $3 for a 12-hour day on the rigs. A new town, Oil City, boomed into a haven for prostitutes, saloon owners, gamblers, and businessmen trying to make a deal. Brothels, bars, and gambling dens sprang open to amuse men who, while braving humid heat, mosquitoes, snakes, and quicksand, had to sleep on the ground where they could. A tent or tar-paper shack was rental property.

Law enforcement became impossible as train loads of thieves, outlaws, and undesirables poured into the steaming area. Mere fistfights hardly stopped onlookers in the muddy streets. Robbery and murder were rampant. In one instance, an ex-Texas Ranger was hired by Trees Oil Company to help keep the peace. This experiment ended when the lawman quelled a miniriot by shooting several of the participants.

Oddly, the frenzy was confined to Louisiana. Caddo Lake's Texas side had almost no play at all. For some reason, oil was generally east of the state line, which split the lake in half.

In spite of the turmoil, so many oil wells were sunk that one day there were no more available land sites. Faced with this reality, the Caddo Levee Board creatively requested bids for leases on the bottom of the lake itself.

Gulf Oil Corporation had a vested interest in the Caddo field. Its first well there had been completed in 12 days at a depth of 800 feet.

H.A. Melat, the area drilling superintendent, and his colleague, W.B. Pyron, met to discuss how they might handle working underwater. Melat came up with a plan, and Pyron passed it along to Frank Leovy and W.L. Mellon at Gulf's Pittsburgh headquarters. Pyron also recommended leasing as much of the lake bottom as possible. Mellon was skeptical and held back his response.

On the day of the lease auction, Pyron did not know what to do. He had no permission to proceed. Even so, because he was a man of action, he went to the auction, in case word came from headquarters. Fifteen minutes before bidding was to begin, Mellon contacted him by telephone to say no. Drilling over water was too uncertain.

Pyron's fervent assurances succeeded. Mellon was won over and gave his approval as the bidding actually commenced. Pyron raced back into the bidding office just in time, and Gulf obtained the rights to 8,000 acres covered by several feet of water.

Melat then oversaw the cutting of cypress trees from the

surrounding forests and had them driven into the lake bottom at the selected site. The pilings served as a foundation for a walkway, pipe rack, and a rough floor. The derrick, also of wood, was constructed on the floor as if it were on dry land.

In May 1911, Ferry Lake No. 1, the first well to be drilled in inland waters, hit. The celebration was loud and long. In 1950, 40 years later, Gulf had drilled more than 275 wells in Caddo Lake and produced over 13.5 million barrels of oil from the property—a good return on the insistent Pyron's first lease purchase price of $30,000.

Lessons learned on the waters of Caddo Lake made shallow-water marine drilling routine. And for the next decade, few sought new refinements in the field.

The search for fresh sources of petroleum, however, was another matter. Global usage of fuel oil, gasoline, and other related products grew at a frantic pace, abetted by World War I and its aftermath. Petroleum prospecting became an international business, and new fields were opened without regard for the borders of sovereign states.

One highly successful exploration company was Shell Oil. In 1917 they daringly sank a small well in northwestern Venezuela on the shores of Lake Maracaibo.

A hundred kilometers wide, a hundred miles long, and a hundred feet deep, Lake Maracaibo lies east of the Sierra de Perijá mountain range and connects with the sea through a narrow opening. Fed by rivers, the shallow southern part of the lake is fresh while the northern extreme is brackish. The Shell exploratory well gave the company cause for hope, and they developed a string of wells while gaining leases on additional lands. Then, in December 1922, all hell broke loose. Shell's Los Barrosos No. 2 came in wild, producing approximately 100,000 barrels a day. That kind of production started a land rush, with Shell in the lead by gaining extensive concessions on the eastern lakeshore. Which left the lake bottom for latecomers.

The hard-won knowledge of drilling over water—the same techniques used on Caddo—was brought south. Felled trees became pilings which were driven into the lake bed to make a foundation for wooden platforms and derricks. Maracaibo was a relatively smooth body of water with no tidal problems. Storms, called "chubascos," came during the rainy season with high winds, but rig platforms could be made sturdy enough to bear those occasional added loads.

Nature had another trick in store, though, that was even more devastating. Its name was "teredo." Teredos, or shipworms, which in the tropical waters of Maracaibo could grow five or more feet in length, had been the curse of mariners since time began. By boring into and eating away at wood, teredos wreaked havoc on the oil platforms, shortening the useful life of each installation to a few months. After futile attempts to find local trees that could withstand teredo attacks, the oil prospectors turned to creosote-impregnated pine timbers imported from the U.S. The costs of building a platform soared with the added expense.

Since those who search for oil are resilient, resourceful individuals, it wasn't long before a teredo-proof solution was developed. Pilings made of steel-reinforced concrete allowed Maracaibo to become the most outstanding oil production area in all the Americas. During a 30-year period, the bottom of the lake yielded more than four billion barrels of oil, and more is still coming.

From that day to this, marine drilling has kept pace with the need for new sources of petroleum. And stalwart people have devoted their lives to conquering the challenge of drilling in deeper and deeper water.

Why this need to work in our oceans?

Viewing our planet from the vantage point of space reveals a great blue ball with wispy stripes of white clouds. One look clearly shows that the name we have given our home, Earth, is inaccurate. Nearly 70 percent of our planet's surface is covered by water, so dry land is the exception, not the rule.

The seas are our future. The mineral riches lying beneath the waves are incalculable. Harvesting that wealth is our destiny.

What does it take to reach down thousands of feet to bore a hole several miles deep into the ocean floor? It takes hard work, advanced technology, and a lot of money. Drilling at today's depths, in our present environment and economy, presents far greater challenges than those met by the redoubtable Colonel Drake.

Anyone who has watched a good rig crew "trip" a string of drill pipe out of the hole understands. As the pipe comes up, big droplets of brownish drilling mud fly through the air, splattering into dirty splotches on every surface. Faces tight and streaming with sweat, even in cold weather, men work at a relentless pace, breaking every third pipe joint apart with huge tongs. Above them, on a narrow platform called a "monkey board," a derrickman helps guide the 90-foot-long stands of drill pipe into racks where they will be stored until time for tripping back down again.

Working day and night, summer and winter, in rain, hail, sleet, fog, or snow, under a sun so hot that metal surfaces can burn skin, or in sub-zero temperatures which can freeze unprotected flesh, these workers do their job. Drilling is tough enough on dry land. On the water, it's even harder.

Offshore, workers labor far from home and loved ones, 12 hours on and 12 hours off, for two or even four weeks at a time, "making hole." Their work base, which can be floating on the water or standing on the ocean floor, is a speck lost in a vast immensity of blue sea. Their environment can change from calm swells and zephyr breezes to the roaring savagery of a hurricane, complete with 60-foot waves and winds up to 17 on the Beaufort scale, in a matter of hours. On top of all that, as they sink their drill bit into the unknown interior of the earth, a sudden pocket of pressure can cause a kick, which if not brought under control in time, may lead to a blowout. Broken lengths of drill pipe can shoot skyward out of the hole, and the rig itself might become a gigantic, blazing torch.

The rig crew, along with caterers, helicopter pilots, supply boat hands, well-service teams, ship's officers, able-bodied seamen, ballast-control specialists, company men, toolpushers, geologists, office personnel, divers, electricians, mechanics, engineers, and many others are needed to make a single offshore well possible.

Unlike those who go down to the sea in ships, the drillers' task is not focused on the wealth in the ocean itself. They are there to probe beneath the seafloor, seeking new supplies of that old substance, rock oil.

Supporting their efforts is not inexpensive. A single well may easily cost $20 million or more. And at the moment the first drill bit is lowered to start opening hole, no one knows if the hydrocarbons it might unlock will be sufficient to make the investment worthwhile. In short, every well, even in proven fields, is to some degree a roll of the dice.

For more than a century, companies and individuals have been bellying up to the bar to make that bet. Why? For some, it's the lure of risktaking; for others, the sheer excitement. For most, it is the opportunity to earn a profit. Or make a fortune. And for many, it's also because they understand the role petroleum plays in shaping our lives.

That's the way it has been since the beginning. And the end is not in sight.

Chapter 3

Off the Land and into the Water

For a time, the Caddo Lake oil boom diverted attention from many other productive Louisiana sites. Throughout the southern part of the state, salt domes were common. A salt dome is a geological formation caused by an upthrusting of rock salt from far beneath the ground. The upward pressure causes a smooth, rounded bulge on the surface of the land which may be many miles in diameter. The rising salt mushroom forces the adjacent layers of sediment to bend upward. Since oil floats on water, even water trapped in sandstone, the sloping nature of the strata around the deeply situated salt plug frequently holds accumulations of petroleum.

The 1901 success of Captain Anthony Lucas with his gusher at Spindletop near Beaumont, Texas, which was not that far from the Louisiana line, caused new interest in salt dome formations. A few years after Spindletop, the Heywood No. 1 well, situated near Breaux Bridge in St. Martin Parish, became the first big producer that demonstrated what is known as the "salt dome overhang." If the upthrusting salt column is bigger at the bottom and top than in the middle, a probing drill on the edge of the formation will encounter salt, then leave the salt for other strata, then hit the lower bulge of salt. Armed with a better understanding of how to tap a salt dome, the search for more of these geological formations was on. Oil prospectors had at last found a way to spot possible petroleum resources in places where no oil or gas seeped to the surface. Other salt domes were located, and even though success was sparse at first, many proved to be effective indicators of profitable drilling sites.

Much of southern Louisiana, however, was bayou, swamp, and marshland, where water obscured the surface, hiding salt dome features. Early use of seismography, which allowed trained engineers to peer into the earth and outline formations by sending shock or sound waves through the ground, helped resolve that problem.

On dry land along the Gulf Coast of Texas and Louisiana, performing a seismic survey came to be called "doodlebugging." The doodlebug is a small, many-legged insect which burrows a funnel-shaped hole into soft earth to make a trap for ants. Its digging leaves a distinctive mound of loose dirt around the funnel. When seismic teams bored openings into the ground to bury the explosive charges which produce shockwaves, circular mounds of displaced dirt formed around the bores, resembling gigantic doodlebug mounds. The portable drilling machines seismologists used to plant their charges were called "doodlebug rigs." Even though the mounds did not show when charting swamps and bayous underwater, the slang stuck. And doodlebugging helped discover one highly productive pool after another.

Innovations made to deal with drilling along the Louisiana waterways opened the door to exploration in the Gulf of Mexico itself.

Drilling in the wetlands was made possible by the use of timber "mats" hammered together out of boards like a giant lattice. Placed on a soft mud bottom in a foot or two of water, they supported the derrick and other hardware. Heavy equipment was needed to drill the deep holes required to probe salt domes, and this special support kept gear from sinking into the muck.

In deeper water, drillers at first followed the tried-and-proven platform-on-pilings system. Then, as exploration lured them farther and farther into the wilderness, another solution was needed.

The challenge was simple. From Louisiana's Calcasieu Lake to Breton Sound, the country was neither land nor water. One old-timer described it as "too thick to drink, too thin to plow." There were no roads, no bridges, and no towns to support drilling crews. People, pile drivers, and the drilling rig had to be hauled by boat or barge, which was expensive enough. Facing the cost of constructing a mat or planting as many as 250 pilings into the soft mud to support a single platform, and then using more than 35,000 board feet of lumber for the platform itself in order to drill what could possibly be a dry hole was too much.

In the late 1920s the Texas Company leased huge chunks of marsh, swamp, bayou, lake, and even Gulf of Mexico shoreline from the state of Louisiana. That enormous financial investment, and a promise of even greater return, spurred them to find a better way to drill in the wet wastelands.

After a few years of study, G.I. McBride submitted an idea for a sinkable barge. Equipped with a derrick and everything needed to drill, the barge would be towed to the well site. Then it would be sunk on the bottom of the swamp to provide the needed stable work platform. When the job was completed, the barge would be floated again, ready for transport to another location.

The idea was warmly accepted. In preparation for building this revolutionary new device, Texas Company attorneys made a routine search of U.S. Patent Office records. The result was astounding. Four years earlier, in 1928, a man named Louis Giliasso, who styled himself a captain, had patented a submersible drilling barge! An examination of patent documents showed that Giliasso had perfectly described the Texas Company plan.

More convinced than ever that the idea would work, Texas Company engineers wanted to begin building their version of the barge immediately. The problem was that Captain Giliasso could not be found. Somewhat discouraged, Texas Company attorneys contacted the lawyer Giliasso had used to file his patent, requesting permission to begin work. Approval was granted, with one proviso: the Texas Company was to furnish a complete set of plans for the finished barge.

A shipbuilding company in Pennsylvania was then chosen to construct the new vessel. While this was being done, a worldwide manhunt for Captain Giliasso ensued. Months later, he was discovered running a bar in Colon, Panama. After selling the bar, Giliasso sailed for the United States.

A native of Italy, Giliasso had been a Merchant Marine captain and had served as a lieutenant in the U.S. Navy Reserve on a troop transport during World War I. After the war, he migrated to the oil business and spent seven years as marine superintendent with the Mexican Seaboard Company at Point Lobos, Mexico. In 1927 he visited Lake Maracaibo and was shocked by the waste associated with drilling on water. After using his maritime experience to concoct a better idea, he drew up plans for his sunken barge, patented it in the U.S. and Venezuela, then tried to peddle his concept to major oil companies.

His efforts did not succeed. The decision makers pooh-poohed his invention. One strong point of resistance was the belief that once the barge sank in the mud on the bottom, suction would hold it there and it would never resurface. Regardless of Giliasso's background in undersea salvage, the oil company "experts" knew better.

So Captain Giliasso gave up his plan, ran his bar, and tended to other matters. News that the Texas Company had been trying to locate him because of his drilling barge concept was exciting. So he hurriedly took a ship to the U.S. Arriving in 1933, he signed an exclusive agreement with the Texas Company to build and use the barge as well as license the idea to other companies.

The completed revolutionary vessel, named the *Giliasso* in honor of the captain, was floated down the Ohio to the Mississippi, then to Lake Pelto, Terrebonne Parish, Louisiana. On a warm, mid-November day, her seacocks were opened and she settled gently to the muddy bottom.

The first well drilled from the *Giliasso* went down 5,700 feet. It was a dry hole, which somewhat negated the enthusiasm for the new technology. When the moment came to refloat the barge, those on board were nervous. After all, some of the best minds said it would be stuck in the muck forever.

Captain Giliasso, however, was vindicated. As air was pumped into its flotation chambers, his invention rose to the surface effortlessly. Afloat again, she was moved 200 yards to a second drill site and the process started over.

The Texas Company had spent roughly $30,000 for the barge. Savings of $75,000 were realized in the first year alone.

True offshore drilling, however, out of sight of land, was beyond technical reach well into the late 1930s. Tides, waves, wind, storms, and logistics made Gulf of Mexico waters too unpredictable. Yet petroleum was thought to be out there in staggering quantities. By one estimate made in the 1950s, near-shore reserves in the Gulf might equal seven-tenths of all gas and oil found in the U.S. up to that time. There had to be a way to unlock this potential fortune.

A joint venture between Shell and Pure Oil made the pioneer attempt in 1937 at a site a mile off the Louisiana shore and about 13 miles from the tiny hamlet of Cameron. Back to the old pilings-and-platform method, a rig and ancillary equipment were installed in 14 feet of water with the main drilling floor another 15 feet above the surface.

This location demonstrated the complexity of logistical support that all offshore operations would require. There was no radio on the rig, so if a breakdown occurred, the next boat to Cameron carried the message and ordered parts. Since Cameron was connected to Hackberry by a single telephone line, which worked about as often as it failed, long delays were common. And because no provisions had been made for men to live on the rig, the crews had a 13-mile run, usually in a rented shrimp boat, over seas that were rarely calm. Facing four to five hours on a seasick round trip every day made already rough work even worse.

Nevertheless, the men endured and their efforts were rewarded. The Shell-Pure Gulf of Mexico State No. 1 well opened the path for drilling in the relatively shallow Gulf waters. It was an exciting breakthrough because the gently sloping continental shelf tapered outward for a distance of up to 120 miles from some places on the Texas and Louisiana coastline.

The pressures of World War II stymied further marine exploration in the Gulf. When the war ended in 1945, prospects for a resumption of activities looked bright. But then came a fight with the federal government over ownership of the submerged acreage. Concern over this issue was especially keen in California, as the question had first arisen there. Along that beautiful coast, where mountains come down to the sea, the floor of the Pacific Ocean contained vast petroleum deposits. Platform drilling, though, was limited to a few dozen yards from the beach. In many places depths of 600 or more feet occur less than a mile from shore. Furthermore, a powerful sea action from waves and swells strains platforms even close to land. And in southern California, where several beaches had been defaced by a maze of derricks extending to the water's edge and then platforms running out in the surf, public sentiment was firmly against this sort of marring sight.

The solution was available for everyone to see. Ships came and went into the great ports. Ships maneuvered on the water and could travel from place to place. So the obvious answer was a ship like none built before—a ship that contained a complete drilling rig and all ancillary gear, including a mud-handling system, drill pipe, and the ability to perform well cementing. Passenger and cargo ships already provided shelter and meals for the crew, so this new tool would have to house roughnecks, roustabouts, drillers, geologists, and the myriad of people needed to keep a bit turning and making hole.

No one had such a grand vision in those early postwar years. There was, however, a seemingly unrelated series of events taking place that would bring together a unique group of men. Together they would earn a place in history by making the drillship a reality.

While the existence of petroleum deposits off the California coast was a proven and accepted fact, little was known about the extent or locations of those reserves. To learn more, it would be necessary to conduct a geological evaluation of a very large segment of the ocean floor.

The Union Oil Company of California had a long history of using the most up-to-date technology in its search for new finds. As early as 1919, for example, the company hired two ex-officers who had served with the American Expeditionary Force in France. Their job was to make aerial photographs of selected California locations which could be pieced together into a map.

This gave company experts an otherwise unobtainable view of the landscape and an opportunity to examine the interrelationship of various ground features.

Union Oil geologists were also armed with tools which had been developed in one world war and improved during the next.

Invented by a Hungarian prior to 1914, the torsion balance had been used by the Germans trying to increase production in the Rumanian Polesti oil fields. Properly handled, the device could detect minute variations in gravity as it was moved across the earth's surface. Changes in gravitational strength indicated alterations in the earth's subsurface density, which helped identify likely sites to drill.

The magnetometer was another greatly refined tool just being harnessed for oil exploration. This instrument plots modifications in the earth's magnetic field and gives another view into subterranean depths.

Likewise, the seismograph, originally designed to record earthquake shocks, had been upgraded. Used by the German army to locate Allied artillery batteries, it was first applied to oil exploration during the 1930s. Reflective seismography developed rapidly, and within a few decades became the industry's most reliable and relied-upon means of looking underground.

Still more direct knowledge could be gained through taking samples of the earth from various depths and studying the sediment for tiny fossilized remains. These samples, called "cores," provided a means of identifying the type and age of the strata in which they were found.

With all of these techniques available, surveying hundreds of square miles of seabed would require a long and expensive effort. The potential return on that investment, however, was staggering.

In 1946, to share costs, Union Oil banded with another major player to undertake the study as a joint effort. The Union Oil alliance with Continental Oil Company (Conoco) was to be staffed by geologists and engineers. Working as a team, these men began a systematic study of the ocean bottom.

Their efforts must have shown a marked degree of promise because, two years later, Superior Oil Company bought into the group, agreeing to a proportional share of all past and future outlays. A year after, Shell Oil Company followed the same path to become the fourth participant.

Continental, Union, Shell, and Superior Oil Companies decided to call their growing entity the "CUSS Group," an acronym derived from the first letter of each firm's name. Their mission remained the same. CUSS was to define the prospects for oil off the California coast.

Headed by Louis N. Waterfall, a Union Oil Company geologist, the small, handpicked group set to their task. First on the agenda was collecting geophysical samples.

The team quickly devised ways to pierce the seafloor and recover cores or small-diameter cylinders of the subsurface earth, which revealed the sequence of strata or layers under the ground. After initial attempts, they realized the cores they had been getting, as one member of the group put it, were "toothpaste." They needed to push through the layers of muddy sediment that covered the seafloor into the bed of the ocean itself. So they began laying out plans for future study.

In 1951 the CUSS Group purchased a 175-foot-long, wooden-hulled U.S. Navy patrol boat. The ship was retrofitted at the Harbor Boat Yard on Terminal Island, California, and rechristened the *M/V Submarex*. The name was derived by combining "submarine" and "exploration."

Experimenting with techniques to collect better cores occupied a number of months and went from scooping the bottom with steel buckets to "drop coring" with a "cookie cutter." The cookie cutter was a heavy steel bar suspended by a wire rope. One end of the bar was equipped with a sharp, metal core barrel. When dropped or thrust downward over the side, the weight of the bar stabbed the core barrel through the sediment into the area to be sampled. Results were mixed, ranging from no recovery to bringing up a core three inches or less in length. The system clearly had room for improvement.

After much work, the team refined their solution. Water

pumped through a length of pipe pressed against the ocean bottom washed away the mud, allowing the pipe to penetrate into the desired zones. Although somewhat better, this method of "jet coring" still fell short of their goal.

During 1952 events in Washington, D.C., indicated the prolonged legal battle over ownership of offshore coastal oil deposits would soon be decided in favor of the states. General sentiment was that California would be quick to take advantage of such a decision and hold auctions for offshore leases. With this incentive, the oil companies owning the CUSS Group decided to expand the scope of their joint venture. To obtain better seafloor samples from depths deeper than those gained by punch core or waterjet methods, they began an investigation of two very different techniques.

The first was to use a drilling rig mounted on a floating platform. The second was to use a deepwater fixed platform to support the necessary men and equipment.

In 1953, with these projects approved and funded, the CUSS Group staff was expanded to add an increased engineering capability. Robert F. (Bob) Bauer was transferred from Union Oil to become manager of offshore operations.

Tall and barrel-chested, Bauer had been a cowboy in his youth, working on a ranch in Utah. While still in his teens, he rode to California on a truck carrying chickens to see his uncle who was in the oil business. The uncle grudgingly gave him a job as a roustabout on a rig. Bauer paid for his college education by working in the oil patch and in 1942 earned a degree in petroleum engineering from the University of Southern California. Coming out of school, he opted for experience over graduate studies and found employment with Union Oil. He was a serious man with a friendly, direct manner and made no secret of his driving desire to have his own company.

Union Oil also assigned A.J.Field and Hal Stratton to support the CUSS team.

Field did his undergraduate work at Cal Tech and received a Master's Degree from Stanford University. He joined Union Oil in 1947 as a roustabout after serving with the Seabees during

41

World War II in the Pacific Theater of Operations. He rose rapidly in the company, holding engineering and supervisory positions. In the offshore group he assumed responsibility for engineering programs and economics. Possessed of a sharp, organized mind and a cordial, earnest personality, he had the qualities of leadership.

Hal Stratton graduated from the University of Southern California and also worked for Union Oil as a roustabout. He was promoted to the position of petroleum engineer. In less than four years, he became drilling and production foreman. Moving to the offshore group, he was named offshore drilling superintendent and directed work on floating drilling techniques. Stratton was an outgoing man who combined both a practical and technical background with a congenial manner.

These three men and their close associates formed the nucleus of a team which was about to revolutionize the oil industry.

Even with the added personnel, the CUSS Group had little practical expertise in marine operations. Which was not surprising, considering that almost all oil fields were at that time on dry land. To fill this gap, it was decided the CUSS Group would hire outside specialists.

Stone & Webster, a firm from Boston, was selected to provide the necessary engineering services. To bolster their skills and to make California operations more effective, Stone & Webster sought a partner. They approached Macco Corporation, a company well known for its work in construction as well as pipelines, derricks, and other oil field-related activities. The result was a Stone & Webster/Macco joint venture to aid the CUSS Group by providing needed technical assistance.

Russell B. Thornburg attended Yale in the NROTC program and graduated from Stanford with a Bachelor of Science degree in engineering. His father had been an oil consultant, and Thornburg had lived in Arabia and London. Dynamic and outgoing, he was the Stone & Webster choice to serve as that firm's representative for projects with the CUSS Group.

A young mechanical engineer, R. Curtis Crooke, who had graduated from the University of California in 1949 with an interest in hydrodynamics and marine engineering, was employed by Macco. He had previously worked on an Office of Naval Research project dealing with ocean waves. In addition to wave forecasting, the effort also examined the effects of wave forces on piles. That background gained Crooke an assignment by Macco to its undertaking with Stone & Webster where he would work closely with Thornburg.

Along the California coast in the early 1950s, deep water was considered to be about 200 feet. It was generally accepted in the industry that attempting to drill and produce oil at any greater depth was not economically or even technically feasible. So the CUSS Group's continued geological survey work was limited to the 200-foot mark.

As part of the fixed-position platform research, they turned to a patented tripod which had been deployed off the U.S. East Coast. Mounted far from land to support electronic gear used in the early warning radar system, this device seemed to have promise. The CUSS Group obtained the rights to work with the design. They made several engineering modifications and produced models that could be towed out to sea and set in water up to 600 feet deep.

At the same time, a floating rig was being considered. A light-duty drilling rig had been mounted on the *Submarex*. The performance of this setup was less than satisfactory, so the team sought to improve it. After removing the original installation, they engineered a gimbal-mounted rotary table. This was subsequently cantilevered out over the water from the port side of the vessel. On this platform was erected a coring rig made up from components used on land. Today such a solution seems primitive. Back then, however, Bob Bauer and his team were inventing the concepts that made today's accomplishments possible.

The *Submarex*, which was half as long as a football field, had a beam of only 24 feet. The ship was powered by two nine-cylinder diesel engines and was capable of speeds up to 20 knots. She was not only the first drillship in the world, she was, and still is, the world's fastest drillship. When the skinny 2 7/8-inch-diameter tubing they used for drill pipe got stuck in the hole, pulling on it would almost tip the vessel, rocking it from side to side as deep as the gunnels.

Since this was the first ship ever to be used for ocean drilling, no one knew if the idea would work—or what unforseen difficulties would be encountered. To drill at all, they needed some means of bringing drilling fluid and debris from the hole back on board the vessel so the mud could be cleaned and reused. Initially, they tried a recirculating head that returned the mud to the surface through a hose, but excessive hose wear caused them to seek another system.

In the end, this pioneering team essentially worked out the basic offshore technique still in use today. First, they put down a base with guidelines to the surface. Then they landed a wellhead and finally added blowout preventers on the seabed which could be operated from the ship. In the later stages of this effort, they also experimented with a small suspended marine riser to allow mud return and facilitate operations. In short, they developed a working system that could be used in deep water to drill an oil well. (For a quick explanation of methods in use today, see the appendix.)

They also began examining ways to come out of the hole, change bits, or perform down-hole tasks, and then reenter the hole to resume drilling.

The fact that the ship would have to maintain its position without straying more than a few feet from a selected point—despite waves, winds, tidal forces, and currents—was another serious issue. Their development of mooring systems, along with a new understanding of the stresses on the drill pipe, contributed greatly to their ability to operate effectively at sea.

From this proprietary work, the CUSS Group obtained several patents.

The *M/V Submarex* served as a trailblazer for innovative solutions to problems associated with offshore floating drilling. By this time, three other vessels, the *EMS*, the *M/V Decatur*, and the *Julie M.*, had been acquired as additional work boats and support barges.

When success followed success, the CUSS Group began

to see beyond mere coring activities to actually drilling for petroleum and then conducting the activities needed for well completion and production from the deck of a ship.

Work from the *Submarex* also launched two entirely new industries: offshore diving and subsea television.

Abalone divers from the Santa Barbara-Oxnard-Ventura, California, corridor were hired to make underwater repairs to a broken hose or deal with other subsurface malfunctions. The performance of these intrepid men would be remembered later, and they would again be called upon to fulfill a vital role in offshore drilling activities.

Undersea television had first been used by the British during a protracted search for a lost Comet aircraft. They employed large studio-type cameras. The *Submarex* was equipped with smaller commercially available cameras which operated inside waterproof housings. Lowered over the side, these units provided very good underwater pictures with relatively high resolution.

In a matter of a few years, the original sampling effort had evolved into a major research and development program. Those involved provided the techniques needed to explore and produce petroleum from deeply submerged lands anywhere in the world. The CUSS Group was, in short, on the absolute cusp of offshore technology.

While this effort was progressing, a long-sought decision was issued in Washington, D.C. After years of infighting, lobbying, political strong-arm maneuvers, and other tactics, it was time for President Dwight D. Eisenhower to fulfill a campaign promise. He was faced with a decision regarding state or federal ownership of the seafloor off those states fronting on the Atlantic, the Gulf, and the Pacific. Called the Tidelands Act, this decree, which nullified several Supreme Court rulings, was to have a momentous effect on the now-blossoming marine petroleum exploration business.

Chapter 4

The Tidelands Issue:
The Fight for a Fortune in Offshore Oil

Oil and water do not mix. Oil and politics, however, seem made for each other. A demonstration of this affinity can be seen in the great tidelands debate. It became one of the longest, most brutal, and intense political battles ever waged in Washington, D.C. The prize was ownership of the rich petroleum resources off the three United States coastlines. Opposing sides did their best to cloud the true issues with rhetoric, including misuse of the term "tidelands."

Geographers and mariners envision a tideland as that strip of land that lies between lowest and highest coastal tide levels. From the low-water mark, and reaching out to sea three miles, was, in the 1940s, generally considered a nation's accepted international boundary. Beyond that was international waters.

The tidelands debate focused on whether the states or the federal government owned the seafloor between the low-tide line and the three-mile limit.

The dispute had a strange beginning.

When Franklin D. Roosevelt assumed the presidency in 1933, he named Harold Ickes as Secretary of the Interior. Blonde and a little pudgy, Ickes, a Chicago lawyer, had aided Roosevelt in the win over Herbert Hoover and then lobbied for the Cabinet post.

Contemporaries described Ickes as self-righteous and sensitive to any slight, real or envisioned. Others held he was "combative, shrewd, and belligerent." In private, he appeared to enjoy his public nickname, "the old curmudgeon." What he apparently never learned was that Roosevelt often referred to him as "Donald Duck" because of his uncontrolled explosions of temper.

Ickes was a stalwart liberal who supported the New Deal with zeal. And his role as Secretary of the Interior was to make him the first American "oil czar" with government control over oil pricing.

Early in his administration, an application was made for an oil lease inside the three-mile limit off the coast of California. This petition was refused because the Interior Department had traditionally considered those submerged lands to be state, not federal, property.

That position, however, was inconsistent with the "one for all, all for one" attitude of New Deal politics. So in spite of a standing 150-year precedent and 30 previous rulings by the Department of the Interior, Harold Ickes became convinced the federal government, not the individual states, owned the seafloor.

The matter of ocean bottom ownership had not been a serious concern to many people, as there was little of value in question. Oil now made the difference. And the issue, in reality, was whether the states or the federal government would receive the millions of dollars which would come from leasing and royalties.

To secure federal claim, Ickes convinced President Roosevelt that the government should sue California. Such an action would settle the affair through legal proceedings. So in 1935 the U.S. attorney general filed a court action.

From the initial ruling through subsequent appeals, the point was fought through the judicial system. Somewhat defused by World War II, and no longer news compared with bulletins from the front lines, the tidelands topic was temporarily shelved but far from forgotten.

The pot began boiling again when Harry S Truman ascended to the highest office after the death of Franklin Roosevelt in 1945.

The new president, in one of his early actions, quickly

asserted federal claim to the entire continental shelf, the seafloor which extends outward from land to the beginning of the ocean's depths.

With the California case pending before the Supreme Court, Truman's act sparked an instant reaction. Legislators from California, Texas, Louisiana, and other states quickly forged a coalition. In short order, a bill giving states the disputed lands was shoved through both House and Senate.

Truman angrily vetoed the measure, claiming the question of ownership was a matter to be decided through the courts. After an attempt to override the veto failed, both sides vowed to continue fighting.

The Supreme Court did not rule until June 23, 1947. That verdict was unequivocal. The California seafloor was federal, not state, property.

Armed with the decision, additional suits were brought against Texas and Louisiana. Since precedence had been established, the legal route to the Supreme Court was faster. Texas, in an effort to complicate the case, based its position on historical fact. Prior to being admitted to the Union in 1845, the Republic of Texas had been a sovereign and independent nation, not part of the United States or Mexico. As a country, Texas had claimed that its territorial waters extended from the shore outward for a distance of three marine leagues (or roughly nine miles). That boundary was unquestioned when Texas was accepted for statehood.

In December 1950, the Supreme Court denied that argument and found against both of the Gulf Coast states. Legally the problem was resolved.

The fight, however, was only beginning. An act of Congress, passed by both houses and signed by the president, could circumvent the Supreme Court decision. The effort in 1946 to override a presidential veto after passage of such a measure had narrowly failed. Now, two years later, those for state ownership of submerged oil lands believed they could once more force the issue by passing a law through Congress. On this attempt, however, they wished to make it clear that a veto would not stand.

They needed an overwhelming majority approval of the bill.

To prevent such a one-sided vote, the administration and those against the measure introduced a compromise. Their substitute act recognized federal jurisdiction and therefore federal development of the tidelands. Income from mineral leases, however, would be divided between the states and federal government, with the states receiving 90 percent of all revenues. Those favoring outright state ownership were prepared for this tactic and handily beat back the proposal.

Debate in the House on the real measure was short. Advocates devoted most of their arguments to criticizing the Supreme Court for further eroding the powers of Congress and decrying the intrusion into states rights. Opponents bemoaned what they termed a giveaway of oil to the big companies and the selfishness of the oil-rich states which did not wish to share the wealth.

On April 30, 1948, by a House vote of 257 to 29, a bill setting aside the High Court's ruling and restoring tidelands ownership to the states was passed and sent to the Senate. The margin of victory was 66 votes more than the two-thirds majority needed to cancel the anticipated presidential veto.

The next battle was in the Senate, where approval seemed certain. Chances for a huge plurality, however, were less sure. The Senate version of the bill was purposely delayed to allow time for more arm-twisting, which was without success. Passage of the measure was gained, but by a slim margin.

Balloting for the final draft of the bill to be sent to Truman showed little change. The House gave approval with 23 votes more than needed to override a veto. The Senate count was 7 short of the required number. The administration had been successful, denying the measure's backers the overwhelming support they had hoped to achieve.

When "Give 'em Hell Harry" Truman believed in a position, he could be totally obstinate. And no one with any knowledge of the subject ever doubted his ability as a tough, pragmatic master of the political system.

His first tactic had been delay, slowing the measure wherever possible during its incubation and hatching. He stayed with what had worked. Finally, on May 29, 1952, time was about to run out. He placed his veto on the bill.

The veto signaled the beginning of a bare-knuckle brawl. Those senators favoring the measure had access to ample funds that could be contributed to the campaigns of other senators who held no strong feelings on the matter. They also enjoyed robust assistance from the states with a vested interest in tidelands ownership. So they had little problem in holding their coalition together.

The administration countered with support on particular projects that were dear to this or that senator's heart and created a formidable opposition. Both groups then went public with their arguments, which quickly degenerated into name-calling.

In the end, Truman's veto stood. He had won the battle. The war, however, continued.

Those favoring state ownership of the tidelands were forced to look to the future. A presidential veto had stopped them twice. So they needed a president who would approve the measure if they could again muster the strength to place a bill on his desk. Who could they find to side with their cause and win?

An impressive behind-the-scenes player had an answer. Sid Richardson, a personal friend of Franklin Roosevelt, grew up around the town of Athens, Texas. A lifelong bachelor, stocky, with thick shoulders, he was a product of the early oil patch.

Entering the business as a well equipment and supply salesman, he became a producer in Fort Worth, Texas, during 1919 and was a millionaire after his big strike came in the Keystone oil field.

Called the first billionaire west of the Mississippi, he had the ability to put people at ease. He seldom carried money and loved to eat in small Mexican restaurants or rib joints.

Since Truman was not going to run for reelection in 1952, the race was completely open. Richardson wanted Dwight David Eisenhower. In typical fashion, Sid sailed on the *Queen Mary*. Once in Europe, he flew to Paris for a conference with Eisenhower, who was winding down his military career.

Richardson had known Ike since 1941 when they met by

chance a few days after Pearl Harbor. Eisenhower had been a colonel, flying on a military plane that made a forced landing. He continued his trip to Washington, D.C., by train. While Eisenhower's berth was being prepared, the two shared a compartment.

Sid wanted Ike to run as a Democrat and replace Truman. But he assured Eisenhower of full backing even if he entered the race as a Republican.

Ike declared as a Republican candidate and Richardson lived up to his word. He, and others who were normally counted in the Democratic camp, came over.

Ownership of tidelands oil became a major debate issue in the hot presidential contest. Adlai E. Stevenson, the Democrat who opposed Eisenhower, stoutly stood for federal possession. Ike was now clearly for the states. He won Texas and the election. His victory appeared to make the tidelands a slam-dunk, done deal.

Truman, however, had other ideas.

During World War II, the executive order, a long-established means for the chief executive to conduct business of state, came into more frequent and pointed use. It was a weapon ideally suited to lock up tidelands oil for the federal government.

On January 16, 1953, in one of his last official acts, Truman issued an executive order declaring all submerged lands to the edge of the continental shelf a naval petroleum reserve. This maneuver gave the Secretary of the Navy administrative control over all the petroleum deposits in the reserve.

The original concept behind the naval oil reserve plan had been to ensure U.S. Navy vessels adequate access to the oil they needed for operations. Since sea transportation routes might be blocked by enemy intervention, three principal petroleum-rich sites had been established.

Upon receiving notice of Truman's stance, screams of anguish issued from those supporting state ownership. They took quick congressional action, realizing that simply rescinding Truman's executive order was not enough. The Supreme Court decision would still stand.

So they rallied for yet another fight.

After much haggling, the House of Representatives passed and sent to the Senate a measure which would demolish the Supreme Court decision. Compromises were introduced, and five and a half months later, the Senate Interior and Insular Affairs Committee was deadlocked. "Delay piled upon delay" again became the opposition's strategy.

A great deal of politicking also went on. Representative John F. Kennedy publicly proposed federal ownership with monies from leasing going to health issues. Others came forward with suggestions that all income be used for education.

Finally, Senator Tom Connally of Texas, contentious and exasperated, introduced a resolution to discharge the Interior Affairs Committee from further consideration of the tidelands issue. Under this threat, a compromise was reached. And new ways were found to delay matters even further.

What ensued was one of the longest senatorial debates in American history. Senator Wayne Morse of Oregon set a new record for time spent filibustering. Standing with his feet planted firmly on the floor, to meet a Senate requirement, he began to speak at 11:40 A.M. as part of a move by 25 Democratic senators to talk both versions of the tidelands bill down. In an attempt to counter this ploy, Senate Majority Leader Robert A. Taft had kept the Senate in session for four nights to physically weaken the opposition.

Speaking slowly and occasionally sipping tea, Morse ranted against the "giveaway" for 22 hours and 26 minutes without stopping. At 10:06 Saturday morning, he heartily welcomed senators who had gone home to sleep and stopped talking because "I have no more to say."

The acrimonious debate raged on until Taft managed to bring the bill to the floor in mid-May. The three-mile limit version passed by 56 to 35. The count was short of the majority needed to overturn a veto, but this time Ike was in the White House.

On May 22, 1953, in a session at Camp David, with more than 40 senators and representatives in attendance, President

Dwight D. Eisenhower signed a version of the bill which gave states title to offshore lands within their historic boundaries. The bill was the first major legislative measure to be enacted by the new administration.

While congressional debate ended, several states took legal action to block the newly approved measure. The question returned to the Supreme Court for the 1958 October term. This time, the High Court found for Texas, Louisiana, Mississippi, Alabama, and Florida.

One of the most interesting elements of the tidelands matter is that, contrary to legislative rhetoric which shaped popular thought, the major oil companies realized little or no gain. The real issue was centered on who would get the millions of dollars which would be paid for leases and royalties. When signing the bill, Eisenhower correctly termed it a states-rights question resolved in favor of the states.

Sid Richardson, who was a publicly invisible player from start to finish, lived to see the bill pass. He died of a heart attack in September 1959 while visiting his private estate on San Jose Island off the coast of Rockport, Texas. At 68 years of age, he was the last of his kind. Burial was held in his hometown of Athens after a memorial service with Billy Graham giving the eulogy.

Answering the tidelands question unleashed what was to become a technological boom in the development of our ability to locate, drill, and produce petroleum deposits from under the waves. Both individuals and companies were to make fortunes in the coming decades. In a very real sense, a race had begun to determine who possessed the ability to drill in deeper and deeper ocean waters.

Chapter 5

CUSS and the World's First Drillship

The tidelands struggle seriously hampered development of offshore petroleum exploration. Marine activity, however, did not come to a complete halt. Even though unsure whom to pay, most of the major oil companies hesitantly made leasing agreements and drilled a limited number of exploratory wells. Even so, the effort was halfhearted because of questions on royalty amounts and the general uncertainty of the situation.

In 1952 the United States produced about seven million barrels of oil per day. Only 70,000 barrels, or approximately one percent, were being recovered from submerged lands. Three-quarters of that came from under the seas off California while the rest was taken in the Gulf of Mexico.

When Eisenhower signed the Tidelands Act in 1953, some $62.5 million in royalty payments had been escrowed and were awaiting distribution to either the states or the federal government. Subsequent court fights held up payment to the states for many months. In the end, California received $47 million, Louisiana approximately $15 million, and Texas realized $500,000. The amounts of those settlements clearly indicate where offshore activity was strongest.

With resolution of the tidelands question, the oil companies which owned the CUSS Group (Continental, Union, Shell, and Superior Oil) saw promise of return on their investment in the offshore geological survey work. Acquiring underwater leases seemed to take on new importance. The proprietary information they had obtained could make them more intelligent bidders. Now the group needed new data about possible prospects—not to mention the ability to drill producing wells at water depths greater than had ever been considered. So they expanded their venture.

The change of emphasis, from determining that petroleum deposits were there to devising a method for sinking a well and producing the oil, caused the CUSS Group owners to reconsider their position regarding CUSS itself. Suddenly, questions of technology ownership, potential liabilities that could be incurred in actual drilling, insurance costs, and similar for-profit business issues became significant. All this was resolved by a single resolution in 1955. The CUSS Group was transformed into a stand-alone corporation named after the man who had been chief of the previous operation. Louis N. Waterfall Inc. was owned by the same oil companies. Waterfall was named president, Bob Bauer was appointed vice-president and treasurer, and the board of directors was composed of one representative from each of the four owners. In its second meeting, the board accepted transfer to the new firm of all outstanding debt of the CUSS Group and title to the four vessels.

The CUSS Group's past success encouraged the four participating firms to press forward. A first step was replacement of the "toy" *Submarex* with a larger, more productive ship.

In 1955 Stone & Webster/Macco located a surplus Navy YFNB manned freight-barge hull in a port on the Gulf of Mexico. Bob Bauer negotiated the purchase from Magnolia Oil Company for under $100,000, and it was towed to Forrester Shipbuilders on Terminal Island in San Pedro Bay, California. While inspecting the new acquisition, Bauer and Hal Stratton found the interior of the barge large enough for a football game. As they played, the vessel rolled so much it made them wonder how they could possibly drill from such a platform.

Doubts, however, did not stand in their way. Bauer, A.J. Field, Stratton, Russell Thornburg, Curtis Crooke, and the rest of their team immediately began installing dry-land drilling equipment on board.

To pay for this conversion, Waterfall Inc., backed by a shareholder commitment to guarantee the note, went shopping for a loan. Finding money was difficult because the concept of a 260-foot, floating, oil well-drilling barge was totally foreign to the financial community.

Nonetheless, in August 1956, Waterfall Inc. received a credit of $2.2 million and was able to pay for work already in progress on the development of their revolutionary vessel. She was to operate in water depths up to 600 feet and be capable of drilling a hole 15,000 feet deep. Meeting that goal required resolution of a series of personnel and technical challenges.

Solving the people problem proved to be simple. Overall, the CUSS team, many of whom had worked on rigs, had a considerable amount of practical, hands-on drilling experience. Even so, they needed a man to concentrate on that aspect of the project and bring to bear the latest available technology. Since they employed captains and crews of seamen to satisfy their maritime needs, it made sense to hire someone to oversee their drilling activities. Leo Gauss, a superintendent for the Santa Fe Company, proved to be their man. Santa Fe had been the land drilling department of Union Oil and was spun off to form an independent organization. Leo came aboard the *CUSS I* and trained many of the drilling personnel.

On the technical side, no one had done such a conversion before. So the engineers handling the huge task of making over the barge were forced to be creative. Their theme became "think your way through it then do it." Brilliant people produced brilliantly original solutions as they took untried ideas and engineered them until they worked. Theirs was more than a cut-and-fit job. It was a complex process involving critical decisions that would affect safety, reliability, and ease of use. Since virtually every task they undertook broke new ground, there were constant questions about compliance with various codes and laws.

Early on, for example, conversion work was hindered by an OSHA (Occupational Safety and Health Administration)-type argument between the state of California safety inspectors and the U.S. Coast Guard. California state law required that all electrical wiring within a given distance of a well site be encased in conduit. Under Coast Guard rules, no conduit was allowed on a ship. The Coast Guard finally prevailed and work continued.

One of the unusual features of what was to become the *CUSS I* was the "moon pool." In order to mount the derrick in the middle of the barge for stability, there had to be a hole through the vessel's bottom for the drill string to pass through. The engineering team enclosed the hole with a 22x32-foot diamond-shaped dam higher than the barge's waterline. This aperture allowed the bit to move downward into the water without ocean flooding in and sinking the vessel. Men on board the *CUSS I* named this area the moon pool because of the radiance of reflected light which made water in the opening glow at night.

The *CUSS I* was to be a complete floating oil well-drilling facility. In fact, she was to make commercial drilling from a floating vessel a reality. Space on board had to be provided for mixing and pumping cement, circulating as well as cleaning the drilling mud, and housing an adequate blowout preventer (BOP) stack. The blowout preventer was a massive set of hydraulically operated emergency shutoff valves assembled in series. If the well being drilled started to blow, the valves could be slammed closed, shutting off the well bore and stopping the escape of oil or gas.

Power for all the necessary equipment was delivered through chain-drive "compounds" or gear boxes driven by diesel engines. One set of chains went below decks to the mud pumps; the other to the drawworks above.

For quartering the various crews, berths were made for 36 men. Amenities included a galley, recreation room, and office space. A compartment was set aside for radio, sonar, and radar equipment.

Derricks used on land were far too feeble to withstand the constant rocking and rolling stresses induced by the ship's motions. So the team drew their own heavy-duty model. Which probably made them the only drilling contractor since the early days of Uncle Billy and Colonel Drake to design their own derrick. With a height of just 98 feet, it was intended to handle "doubles," or two 30-foot joints of drill pipe screwed together. Maximum total weight it could sustain was about 550,000 pounds.

The storage or "racking" of the drill pipe also had to be

rethought. Standing pipe on end, as is normally done on land, presented a severe safety hazard on a pitching ship. The group developed a pneumatic pipe-handling system which racked the doubles automatically in a horizontal position. This operation was console-controlled from the derrick floor.

Part of the pulley system in the derrick also demanded a departure from land operations. The "traveling block," which moves up and down inside the derrick as pipe is raised or lowered, could not be allowed to swing free. The oil companies fought the expense of securing this heavy unit. Money talks, and they won. Once at sea, though, less than 10 minutes of operation convinced all doubters. The immense unrestrained mass of steel pulleys and cable almost wiped out the derrick house. Fortunately, the engineering team had designed the derrick so that it was an easy matter to add bracing, guide rails, and spring bumpers to tie down the traveling block.

Mooring the ship presented another challenge. The engineers purchased the same brand of chain ordered by the U.S. Navy. Ships' anchors, however, are used intermittently. Under continual employment as the connection between the *CUSS I* mooring buoys and a 7,500-pound Danforth-type anchor set into the seafloor, the chain corroded, causing dramatic shortening of useful life. Having a chain suddenly snap, allowing the barge to swing about while the drill string was a thousand feet into the hole, could be disastrous. And the process of replacing a broken chain required hours of concerted effort. Eventually the planners settled on a welded "stud-link" chain, which came to be called "oil-rig quality," a type still in use today.

Six mooring buoys, each chained to a separate anchor, were used to hold the barge in position. Running from three diesel-driven winches on the bow and three similar winches on the stern were six 1,800-foot lengths of 1¼-inch wire line. Each cable ran from a winch to a buoy and back to the barge. Winching different lines allowed the vessel to be placed in a precise spot and held on station as long as necessary.

For drilling under hundreds of feet of water, the engineers devised a system utilizing what they came to call a "birdcage."

The birdcage was a triangulated, tubular-steel support structure designed to rest on the seafloor. It provided a base for wellhead equipment, such as the BOP stack, and carried guidelines to assist in lowering materials from the surface.

Once the barge was securely moored, the birdcage was suspended in the moon pool by wire cables. Lengths of relatively large-diameter pipe, called "surface pipe," were lifted by the derrick and threaded through the birdcage. The pipe was then let down nearly to the ocean floor and allowed to hang just clear of the bottom. The bit and drill string were then lowered through the pipe to spud in the well. Saltwater was used for drilling fluid.

As drilling progressed, a reamer was added above the bit to enlarge the hole to a desired diameter. Lengths were then added to the surface pipe, and it was fed into the opened hole. Once in place, the drill string was removed. Next, an original-design Cameron Type-U blowout preventer package was attached to the birdcage, and the drill string was used to land the entire assembly over the hole on the bottom. After landing, the birdcage, all attached gear, and the surface pipe were cemented in place. Other equipment could be lowered to the bottom through the use of guidelines, running from the birdcage upward into the moon pool. The marine surface or riser pipe was fitted with a buoyancy tank at top, which held the string in tension. Slip joints, basically two lengths of pipe telescoping one inside the other, were used on the drill string and riser line. This provided the vertical free play needed to compensate for the lifting and falling of the barge as it reacted to wave action.

A quick review of the appendix of this book will show that, all in all, this original drilling system clearly pioneered the methods in general use today.

The *CUSS I* also expanded the reliance on divers to assist in offshore oil operations. After World War II, the U.S. Navy Experimental Diving Unit, headquartered in the Washington, D.C., Navy Yard, made great strides in diving techniques. This training and research facility studied the use of special mixtures of breathable gases to allow divers to work in deeper and deeper water. The careful, scientific decompression and time-at-depth tables produced were vital to diver safety, greatly facilitating the growth of this trade.

California divers had their own less formal instruction. Abalone hunters had farmed the ocean around Santa Barbara for decades. The hard-shelled mollusks clung to algal-covered rocky bottoms in waters up to 100 feet deep, forcing divers to work at serious depths. Most of the meat harvested was traditionally shipped to San Francisco and on to Asian markets. So the men who gathered abalone were dealing with a cash crop and treated their underwater activities as a business. Accustomed to operating in a hazardous environment for a living, these divers found the tasks related to oil well drilling not all that daunting.

Lessons learned earlier on the *Submarex* about the value of divers were not forgotten. When the engineers involved with the *CUSS I* found a need for subsurface human activities, the abalone divers who had learned on the *Submarex* formed a ready pool of qualified and willing specialists. As offshore operations grew through the next decades, demand for diver services dramatically increased. And the need for improved abilities to operate at progressively greater depths led to the development of the manned underwater observation submarine and the Remote Operated Vehicle (ROV).

In a very real sense, the *CUSS I* can be credited with furthering this new undersea industry which had been created while working with the *Submarex*. Several of the original divers who assisted on that small vessel went on to become multimillionaires by forming their own companies. They served as leaders in this exciting field that still plays such an important role in deepwater exploration today.

Parallelling another successful *Submarex* innovation, the *CUSS I* was also equipped with underwater TV capabilities. Cameras on board the *CUSS I* were the effective, small Vidicon units with custom-built waterproof housings. Their link to the ship was by special cable laid inside a protective braided-steel hose which was filled with nitrogen gas under light pressure. If the flexible cable covering were nicked or sprang a leak, the exiting gas would keep water from rushing in and ruining the cable. This system worked very well, and along with visual inspections of the well-head and blowout preventers, its many uses became standard procedure.

Considering that no one had ever before converted a barge into a drilling vessel, the work was accomplished in an astonishingly short period of time. In 1956 all systems were ready for testing. The *CUSS I* was a full-fledged floating driller—as well as the most advanced marine drilling tool of her day.

To support the new barge, a small cargo ship, the *M/V Noodnick*, was bought for $31,000. Another $100,000 went for modifying her into a supply-anchor boat. She was rechristened the *Rigger I*. Two of the company's other boats, the *Decatur* and the *EMS*, were sold.

The *CUSS I* was chartered by Union Oil on behalf of the CUSS Group to continue development of methods and equipment for deepwater drilling. In this same year, to better handle long-term lease contracts, Louis N. Waterfall Inc. established a subsidiary called Global Marine Drilling Company. This was the first use of the name which would come to stand for progress and performance. Use of the word "global" clearly indicated those involved had more than an inkling of the potential worldwide demand for their unique technology.

The leasing agreements between Waterfall Inc. and Union Oil were transferred to Global Marine Drilling Company, which became the prime contractor to Union Oil.

In spite of President Eisenhower's signing of the Tidelands Act, drilling off the coast of California within the three-mile limit was still prohibited by state law. The whole matter of offshore petroleum exploration and production had become a political quagmire. Issues of aesthetics and pollution were aired and re-aired during endless debates. Objectors to marine drilling also involved themselves in efforts to prevent leasing offshore federal lands, attempting to completely halt further petroleum exploration.

During the many months between 1953 and 1958, personnel from the supporting oil companies, backed by members of the CUSS staff, then the Waterfall team, began extensive political action work in Sacramento. Bob Bauer's training as a petroleum engineer may have done little to prepare him for a career as a lobbyist, but that did not stop him. He spoke to any group which would listen.

Never one for half measures, Bauer went to a Union Oil executive and solicited the aid of a highly capable company attorney. The two drafted the basis of a bill that would allow marine drilling. To support this position further, Bauer wrote a speech which a key Union Oil corporate officer gave before the state senate.

In the end, and against much opposition, legislation favorable to oil exploration was enacted. There was widespread expectation that leases would be granted. The lapse of still more time with no sign of an offering from the state was disheartening. And expense by expense, the combined investment in deepwater drilling technology grew. By 1958 some $15 million had gone into various CUSS and Waterfall Inc. projects.

With no indication of when the state might relent, the four participating oil companies that owned Waterfall finally reached their limits and moved to end their relationship.

The Waterfall holdings included ships, boats, and a number of potentially valuable patents. So the company had real assets. Which meant it was worth a significant sum of money.

The prospect of dissolving Waterfall Inc. was devastating to those involved in the company—including the staff of Waterfall's engineering consultant, Stone & Webster/Macco. The entire team had given their best and enjoyed numerous successes. Since commissioning, the *CUSS I* had drilled almost five miles of hole in water depths varying from 45 to 350 feet. She was very much a practical reality, and the Waterfall group was on the brink of even greater advances. Many were also about to find themselves unemployed.

As a parting gesture, discussions were held about the possibility of putting together a company that might purchase the Waterfall assets and establish an independent firm which would function on its own. The new entity could then offer the technology to all interested parties in the oil industry.

The key players, Bob Bauer, A.J. Field, and Hal Stratton, still had their old relationships with Union Oil intact. These men essentially remained Union Oil employees and knew the company. Field had been offered an important post, and positions were undoubtedly available for the others in the Union organization or one of its subsidiaries. Accepting those jobs, though, meant splintering the team which had worked so well together.

Bauer's entrepreneurial instincts came to the fore. After gaining permission from Union, he threw himself into financing the new venture. He approached Stone & Webster, then Santa Fe Drilling. For differing reasons, those promising sources were unable to participate. So he turned to Wall Street. No one there said "no." More importantly, no one said "yes" either. At one point, his quest for cash put him in touch with representatives of the Santa Anita horse-racing track. Their talks did not prove successful. Exasperated and disappointed by lack of progress, Bauer opened negotiations with the original CUSS partnering companies.

Through his efforts and determination, Union Oil finally agreed to back this venture. The arrangement called for Union to purchase all outstanding shares of Waterfall Inc. from the other three oil companies. Then Waterfall Inc. could be dissolved and its assets transferred to Global Marine Exploration Company. Union Oil would hold 80 percent of the stock, and the new company's employees would receive the remaining 20 percent.

On November 12, 1958, Global Marine Exploration Company was formed as a for-profit corporation in the state of California. Global Marine Drilling Company remained a subsidiary and served as its parent's key operating organization. Bob Bauer became president. A.J. Field was named vice-president and general manager. Hal Stratton accepted a vice-presidency. According to their agreed division of labor, Bauer would direct the company's business affairs. Field was placed in charge of operations with all employees reporting to him. And Stratton would supervise activities on *CUSS I* and be involved in marketing activities.

The new firm assumed a $3 million indebtedness incurred by the two predecessor companies, and Union Oil produced

$500,000 for operating capital.

Global Marine immediately went into an earnest sales, engineering, and organizational mode. Even so, the deal was risky at best. The talent was there, however. And this team knew more about drilling from a floating platform than any other group in the world. What they didn't know was whether or not someone in the next several months would pay them to put their knowledge to work. Fortunately, the wait for an answer was short.

In late December, after officially having been in business only a few weeks, Global Marine Exploration Company signed its first day-rate drilling contract with Standard Oil Company of California, now Chevron. A "day-rate" deal essentially sets a dollar amount as a basic fee for all time spent drilling a well. That method of compensation is opposed to charging for every foot of well depth drilled.

This display of trust, especially because it did not come from one of the four original CUSS Group partners, gave those running the new business fresh confidence.

Even though work was at hand, operating expenses had to be reduced. As a result, technical assistance from the joint venture between Stone & Webster/Macco was terminated. Curtis Crooke, the engineer who had contributed to the development of the *CUSS I*, established his own firm and shortly had Global Marine as his client. By 1959 he had become a full-time Global Marine employee, again working with Russ Thornburg, who had also joined the company.

Despite management changes within Union Oil, that organization continued to supply Global Marine with operational capital. As agreed when the new corporation was formed, Union Oil increased the number of shares from 10,000 to 500,000. Union retained its 80 percent and, in keeping with its arrangement, offered the remaining 20 percent to Global Marine. There was only one difficulty. In the original plan little mention was made about paying for those shares. Now they were to be available at a price. That news caused some consternation within the new company and resulted in another scramble to raise the needed money.

The year 1959 was notable for another reason as well.

Texaco obtained drilling rights to seabed acreage in 300 feet of water off the California coast near Santa Barbara. Known as the Jade lease, this property was to prove very productive. The Texaco contract with Global Marine set two historic landmarks.

First, the deal called for drilling at least three wells, so it was Global Marine's first multiwell contract. Secondly, the wells became the initial offshore production sites to utilize a subsea "christmas tree" and a connection to shore by underwater pipeline for the produced oil. A christmas tree is a stack of valves, chokes, and pressure gauges used to control the flow of oil or gas out of a wellhead. The assemblage of steel components has more or less the appearance of its namesake and, up to this point, had never before been installed on the ocean bottom.

The field eventually produced from about 20 different wells. And over the next two decades, Global Marine crews, using various vessels, performed scores of maintenance and production-enhancement jobs to help realize the find's full potential.

Global Marine's reputation as a can-do organization spread throughout the oil industry. Major companies like Texaco, Shell, Humble Oil (now Exxon), and SoCal (Standard Oil Company of California, now Chevron), all utilized the *CUSS I* during 1960. The firm was so highly regarded for its technical ability that it was offered lease or purchase deals on several smaller drillships and work vessels which had been developed by oil companies not involved in the original CUSS Group.

SoCal's *Western Explorer*, Humble Oil's *SM-1*, and Richfield's (now Arco) *La Ciencia* and *Rincon* were all brought into the growing Global Marine fleet.

The *SM-1* was in many ways the best of the lot, and in November 1961, shortly after acquisition, she was placed on a job for Texaco off Point Conception in California. A severe storm with long-period swells and a strong southeasterly wind was her undoing. Water was forced into the exhaust pipes, stopping the engines. With no power to run pumps, the *SM-1* slowly filled and all hands watched her sink. No one was injured. Some say, though, that according to the insurance claims, the crew's lockers contained an unusual number of gold wristwatches and fairly large sums of cash.

From the company's earliest days, Global Marine began receiving requests from clients, including the U.S. government, to perform other, often non-oil-related undersea tasks.

A division of U.S. Steel, for example, lost an ore boat, the *Carl D. Bradley*, in Lake Michigan. Traveling in ballast, heading north toward Gary, Indiana, during bad weather, the vessel sank on Thanksgiving night near Middleground Shoals. Of 35 hands on board, only two lived to tell of that awful experience. The hulk lay in more than 600 feet of water, considerably beyond the depth to which divers could descend.

Hearing about Global Marine's pioneer work in the use of underwater TV, representatives of the steel company came to Los Angeles for a demonstration. Cameras were lowered over the side of the *CUSS I*, and good resolution pictures were obtained. In 1959, after a contract was signed, the *Submarex*, which had been laid up, was reactivated at Long Beach, California. With TV equipment loaded aboard, the tall, slender ship began a 7,000-mile voyage to the U.S. East Coast, then north to the St. Lawrence seaway and into Lake Michigan. Once there, the crew located the wreck and performed an underwater-TV site survey. Threatened by the oncoming ice season, the *Submarex* spent several freezing winter months in Charlevoix, Michigan. With the first signs of spring, she was back on site and the crew successfully removed plate segments from the ore barge's hull. Those samples later showed that the Bradley had been weakened by cold, metal fatigue, and age. It was another outstanding, state-of-the-art Global Marine performance.

The *Submarex* was also used to drill an exploratory hole for El Paso Natural Gas in Lake Erie and to recover a number of sediment cores from the bottom of Lake Superior.

Upon returning from the Great Lakes to Long Beach, the *Submarex*, that game ex-Navy patrol boat, was docked. After almost sinking alongside her pier on a couple of occasions, the world's first drillship was sold for scrap in 1965. She had seen her day and done her duty.

Closer to its California home, Global Marine accepted an

offer from the U.S. Navy to install an underwater test facility for the Polaris missile at a depth of 700 feet not far from San Clemente Island. Firestone Tire and Rubber Company also called upon Global Marine to assist with an experimental program for an underwater storage-bag system.

Demand for Global Marine's engineering services had grown and was producing significant revenues for the firm. In response to this, and to improve management of unusual assignments without disrupting the primary business activities, a separate organization was needed. So in 1961 Global Marine Engineering Company was created as a wholly owned subsidiary to administer and work on the diverse projects as well as handle all aspects of U.S. government contracts.

All in all, Global Marine was exerting a tremendous leadership influence in both undersea technology and the field of marine drilling. It was time for the company to take another giant step forward. Using all they had learned, the engineering team began designing the *CUSS II*, an entirely new generation of drillship. Coincidentally, the *CUSS II* was to be built by Equitable Equipment Company, headquartered in New Orleans, just a few miles upriver from the Gulf of Mexico. In those warm waters, a completely different approach to offshore operations had been taking shape.

Chapter 6

Into Gulf Waters

While most of the oil produced from undersea wells during 1953 came from the California coast, savvy petroleum experts eyed the Gulf of Mexico expectantly. Along the Pacific shore, as one Global Marine executive put it, "You could drive a golf ball far enough out to fall into 500 feet of water." It was a different story off Texas and Louisiana. Flat coastal prairies and marshes slid slowly under the waves. The continental shelf was more an extension of land into the Gulf. At the Sabine River mouth, water depths stayed above the 600-foot (100-fathom) mark, which indicated the start of the continental slope, as far as 140 miles from shore.

During the late 1940s, after the end of World War II and prior to the question of tidelands ownership, major oil companies rushed to the edges of the Gulf. Shell, Magnolia, the Texas Company, Kerr-McGee, Pure, Superior, Stanolind, Continental, Phillips Petroleum, Sun, Standard, Texas Eastern—they all came, expecting big plays. The Texas-Louisiana coast was an ideal site for production because the proximity of vast refining capacity and pipeline availability made transportation to market easy. An almost unbroken line of chemical plants and refineries was being erected from Freeport through Texas City, up the Houston Ship Channel, over to Baytown, on to Port Arthur, Orange, Lake Charles, and points east. This was the petrochemical belt, and it had grown on a never-before-seen scale that would continue for years to come.

But the open Gulf posed some daunting challenges.

One was the composition of the bottom. The Gulf seabed was flat enough. Its density changed, however, from spot to spot, so each drilling site might be different. And there had been no systematic Gulf-wide bottom sampling to provide design parameters for platform stability. It was not until after damage was done, for instance, that anyone imagined the huge underwater mud rushes that can accompany storms near the mouth of the Mississippi River. Platforms, pipelines, and other seabed-mounted marine gear were swept away by this powerful Gulf phenomenon.

Those same storms that caused mud flows were only part of the adverse weather problems. From the first of June to the last of October, hurricanes were an annual threat.

Debate raged for years about how high a platform had to be raised above the rolling water to be protected in a hurricane. It was recognized that safe height was directly related to water depth. But there was disagreement as to how many feet it took to be secure.

No one was quite sure about the forces generated by waves and swells, either. Which meant there were no solid data on how strong pilings or platform legs had to be in order to anchor to the bottom and provide stable support for work and living areas above.

A nautical infrastructure was also lacking. Crew boats to haul workers out to rigs, supply boats to carry provisions, barges to transport mud, and vessels with cranes for heavy lifting were either nonexistent or in short supply.

Clearly, much work had to be done. And it would be, in a brief span of years. In the oil business, the term "deep water" simply means water too deep to work in today. It is a constantly changing definition. The urge to drill in deep water has been insatiable.

In the sweltering fall of 1947, just over 10 miles south of Terrebone Parish, Louisiana, in a shallow, submerged area known as ship shoal Block 32, several big oil patch players came together. Almost two years earlier, Kerr-McGee Oil Industries had conducted some rather primitive, by today's standards, seismic testing. Their results identified a pair of salt domes. In 1946 Kerr-McGee paid the Louisiana Mineral Lease Board more than

$30,000 up front for drilling rights to some 40,000 acres under about 18 feet of water. Annual rents would be close to half that amount. To help with the exploration costs, Phillips Petroleum Company and Stanolind Oil & Gas Company bought into the deal. Kerr-McGee became operator of the property. An operator manages the drilling and is in charge of production decisions.

Working against a tight deadline, the Kerr-McGee team ruminated over a cheaper way to drill the offshore site. They could have, as everyone else in the past, built a 30,000-square-foot platform mounted on top of an enormous number of pilings pounded into the bottom of the Gulf. That worked. With a sufficient number of piles, the structure would be strong enough to withstand the efforts of the wind and the sea to destroy it.

There were a pair of major drawbacks, however.

First, the operator was going to drill two wells in different locations. That meant two platforms—at almost twice the expense.

Secondly, if the holes came in dry, the two platforms would have to be dismantled. Government rules required removal of all physical structure down to at least six feet below the surface. The more piles, the more costly the cleanup.

That last fact triggered the engineers' thinking. Smaller fixed platforms, with some of the necessary equipment and supplies carried by a separate, floating tender vessel, had been used in the Louisiana swamps and on Lake Maracaibo. No one had tried the concept out of sight of land in the Gulf of Mexico, where the forces of the sea ruled. Even so, a smaller platform, one just large enough to serve the needs of those drilling the well, could be economically advantageous. There was also danger in the idea.

A huge tender carrying tons of equipment would have to be moored securely so as not to smash into and destroy the platform. In case of a storm, it would be necessary to tow the tender away from the platform and perhaps even place it in a position where it might offer some protection to the unmovable platform.

Kerr-McGee ran tests to determine the strength and size of piling that would be used for support. By driving a 24-inch-diameter steel pile 104 feet into the bottom of the Gulf, satisfactory footing was gained. Then, by careful design, planners devised a 38x71-foot work area which could be supported by 16 steel pilings backed by 6 creosoted wood piles. The roughly 2,700-square-foot surface was painstakingly laid out to make maximum use of space. It would be elevated 33 feet above mean low tide.

The second necessary ingredient was the marine-support equipment. A tender barge would provide living quarters for the men and space to handle mud pumps, mud tanks, along with storage for mud chemicals, cement, logging equipment, pipe, and all other supplies that could not be accommodated on the platform itself. They would also need supply boats and crew boats to transfer men to and from shore.

Uncle Sam's war surplus had just the right vessels at bargain prices. Kerr-McGee bought a pair of 260-foot Navy YF barges; one LST, which was a landing craft used in beach assaults; and three 85-foot air-sea rescue boats.

Both the large barges were gutted, and one was rebuilt as a tender, able to handle enough materials to support drilling a hole 12,000 feet deep. Renamed the *Frank Phillips*, it came complete with crew quarters, galley, hospital, recreation room, and lounge. Home support base for the operation was Berwick, Louisiana, more than 50 miles away at the mouth of the Atchafalaya River. Moving men back and forth daily would be time-consuming. So two crews would live aboard the tender while working 12-hour, seven-day-a-week shifts before returning to shore. Then two fresh crews would be sent out to replace them. The 12-on, 12-off, one-week work period became standard in the offshore industry.

The second barge, leased to Brown & Root, the construction firm which was to install and build the platform, was converted to a seagoing pile-driving and fabrication facility. The LST became a supply boat, and the three smaller craft were used to transport personnel.

On September 9, 1947, all was ready and the initial well was spudded. Thirty-five days later oil was assured, and four weeks after that, the well was complete. Less than 10 percent of the total number of work days were lost when weather forced the barge to be moved out of position. On the second well, lost time dropped below 5 percent.

Kermac-16, as the original miniplatform was called, established a number of firsts: first well to be drilled out of sight of land, first in the Gulf to provide offshore living quarters for crew, and first to prove that the separate platform-tender system was practical in open waters.

Even though effective, the miniplatform and tender method had its drawbacks. Transferring drill pipe from one to the other did not prove easy. It was also problematic to maintain a continuous flow of mud from tender to drilling rig. And the danger inherent in tethering a floating barge to a fixed platform was ever present.

In spite of shortcomings, those in the oil industry were quick to respond to this innovation. The economies were obvious. And total preparation time required to set up a site for drilling was dramatically reduced.

By the fall of 1948 a new industry had been launched. Large-scale conversion of Navy-surplus vessels into tenders was well underway in Gulf shipyards from Mobile, Alabama, to Galveston, Texas. Demand for World War II barges soared, and prices rose sharply. What had been a bargain was suddenly less so. And when the cost of the vessel plus the expense of conversion grew high enough, building a tender from scratch was economically feasible. Custom designs led to self-powered ships, which were more efficient. By the early 1950s, nearly 90 percent of all wells in the Gulf of Mexico were drilled using miniplatforms and tenders.

As popular as this system became, it still left much to be desired. While the small platform could be dismantled from its pilings and moved, the cost of making a platform from these recycled materials was about 60 percent of constructing a brand-new one. And since piles had to be driven, there was a limit to how deep the water could be. There had to be a better way to work in the Gulf to tap petroleum deposits that might lie under a hundred feet of sea.

By the mid-1950s progress in prospecting techniques allowed geologists to operate with new confidence. Swiftly evolving seismic surveying technology was supplemented with improved gravity and magnetic detection methods. While drilling for oil remained a high-risk undertaking, the chances for success were becoming better.

In response to this bonanza of information, the need to drill in deeper Gulf waters grew progressively more apparent. For the first time, the industry was talking about the ability to operate at a depth of 100 feet. Demand was also in place for an improved platform. Ideally, it would be portable, could be transported from site to site, set into place quickly, and be stable enough to avoid weather-related problems.

John T. Hayward had part of the solution.

Hayward had been educated in Liverpool and was a marine engineer in Belfast. He had worked in the Polesti oil fields in Rumania and been supervisor on the first rotary rig in that part of Europe. After coming to America in 1929, he'd served with Barnsdall Oil Company as their chief engineer and head of research. During World War II he was given the Distinguished Civilian Service Award for his antisubmarine efforts.

After the war, Barnsdall, along with some partners, acquired oil leases in Breton Sound, a shallow Gulf area near the Mississippi delta. Having no experience in marine drilling, the group approached an expert for a bid. They had decided six wells would be needed to evaluate the property.

When the partners learned it was going to cost nearly a half-million dollars per well just to construct the platforms and shift equipment, they were flabbergasted. Especially when they realized the half-million did not pay for a single inch of drilling. The hole was extra.

Barnsdall called John Hayward for advice on reducing the horrendous cost. Hayward, instead of reviewing the figures, asked for a little time.

A few months later, Hayward proposed a movable platform which he figured would cost under $300,000, including shifting it from site to site. His solution was based on the idea of a sinkable barge, which would rest on the bottom. Extending upwards from the barge would be tall steel pilings to support a platform that would hold the derrick, drilling equipment, and necessary structures. If the platform were to be 30 feet above the sea in 20 feet of water, the supports had to reach upward from the barge for 50 feet. When the barge was empty of water, it would float and, although ungainly, could be towed. When filled with water ballast, it sank to form a solid base.

The idea was straightforward enough and not unlike Captain Giliasso's earlier barge design on a larger scale. Two questions remained unanswered, however. Would the barge be able to float such a huge mass? And if it did, would the whole rig be stable enough to be towed through choppy Gulf waters without capsizing? There was only one way to find out.

They began building the *Breton Rig 20* at the Alexander shipyards in New Orleans, making drawings and crucial decisions on a day-by-day basis as problems arose. When Barnsdall and partners saw the plans, there was great consternation. A marine engineer was called in, and his comments only made them feel worse. The monstrosity being built would, in the specialist's opinion, not float, turn turtle if it did, and never settle evenly on the bottom.

None of the group knew what to do. They needed to start drilling on the lease, they had an investment in Hayward's folly, and at the same time, they were fearful of throwing good money after bad. All Hayward had was his belief in what he was doing. And most of the experts said he was insane.

One man, the chief engineer at Seaboard Oil Company, partner with Barnsdall in the Breton leases, spoke favorably about the idea. Whether he was reflecting respect for Hayward's proven ability or actually saw potential in the concept is unknown. His opinion, though, swung the group over and Hayward prevailed.

In the early months of 1949 Hayward's *Breton Rig 20* was completed. It was time to see who was right and who was wrong. Three tugs tied lines onto the massive barge and elevated structure. Then they began the tow across 70 miles of open Gulf waters. Winds varied between 20 and 25 miles per hour during the tortuous journey. Everyone involved hoped the new vessel would work as planned. None, including Hayward, was absolutely certain.

Once on site, the sinking process started. Since there had been so many conflicting opinions, they used extra caution. Slowly, the immense steel barge settled deeper into the water, finally coming to rest at a depth of 18 feet, hard against the bottom. Work on the drilling floor began almost immediately.

It took the crews 86 days to make just over 10,900 feet of hole. Then well testing, to better define what was there, required almost three more weeks. During that time on station, the rig remained stable, even in the face of storms with winds of nearly 70 miles per hour.

With the first well done, it was time for the real trial of this new innovation. Would suction hold the vessel tight to the bottom or would it float? Hayward had a trick ready. As pumps began replacing water in the barge with air, the entire rig became lighter. To help break free from the layers of ooze on the bottom, he passed an electric current through the barge and into the water. The charge repelled tiny particles of mud and helped break it loose. In about one hour after giving the order to begin, the vessel was ready to be towed 13 miles to the next drilling site.

Much to everyone's relief, the new platform worked. Before the balance of the wells had been drilled, it had proven its worth.

An accomplishment of this magnitude generally draws a great splash of attention in the oil patch. For some reason, the *Breton Rig 20*, designed in a style which came to be known as a "posted-barge" rig, caused less than a ripple of interest. For almost five years, it was the only movable-platform marine rig in existence. The miniplatform and tender combination was too well established. Operators knew how to work in that mode and were relatively satisfied. Besides, they had large investments in their tender boats. The advantages of the *Breton Rig 20* were obvious. But the financial disadvantage of making an entire fleet obsolete was equally clear.

Kerr-McGee eventually bought the Hayward-Barnsdall *Breton Rig 20* and put it to good use in subsequent years, drilling in waters up to 22 feet deep. They also developed and built a successor, *Rig 44*, which was capable of working in 40 feet of water.

John Hayward, however, was not through in the Gulf of Mexico. He became a consultant and served on the board of directors of Ocean Drilling and Exploration Company (ODECO), an offshore-platform drilling firm. In 1954, shortly after resolution of the tidelands ownership issue, ODECO, headed by Alden J. Laborde, an Annapolis graduate, built *Mr. Charlie*. Derived from Hayward's patented concepts and constructed in the same New Orleans shipyard as the *Breton Rig 20*, *Mr. Charlie* operated at 40-foot depths. This design used fewer columns to raise the work platform above the water and had massive hinged-pontoon extensions port and starboard to add stability.

A year or so later, ODECO created the *John Hayward*. This was a posted-submersible barge with large fixed-hull extensions on the sides and the ability to work in up to 30 feet of Gulf water.

Between 1955 and 1962 a variety of other submersible designs were drawn and built. As this type platform evolved, the barge was replaced by a grid of gigantic tubing which gave a wider, more solid footing on the bottom, yet retained stability during towing. Buoyancy was gained by connecting the grid and the platform with posts which looked like massive steel bottles. A "bottle" is a vertical cylinder which is closed at the bottom and has a top that tapers into a truncated cone. Bottles attach to and support the rig floor. Water can be pumped into and out of the bottles as ballast.

The day of the submersibles, however, was ending. In 1962 Kerr-McGee completed its *Rig 54*, the largest submersible in the world. Able to drill in 175 feet of water, *Rig 54* was triangular in shape, with each side of the triangle measuring 388 feet.

John Hayward's introduction of a movable platform which could be transported from place to place in open sea established a baseline for future technology. The posted-submersible barge concept, though, was costly to construct. And the deeper the water, the bigger the rig. Which drove the price higher. A submersible rig made to park in 60 feet of water with the work area 40 feet above the low-tide line was gigantic because the distance between the barge or grid on the bottom and the platform was fixed. An immense amount of steel and an inordinate number of hours were required to build one of the monsters. And when it was floated, tugs were forced to handle an unwieldy structure more than 100 feet high. No matter how well the submersible was ballasted, safe towing remained a delicate operation.

There was clearly a need for another solution. And in the marine drilling business, needs are seen as opportunities. So an answer was not long in coming.

The new idea was a variable-height work platform.

The DeLong Engineering & Construction Company in New York had done advanced work on a pier, which could be built in a yard, towed to a remote site, such as a location on Venezuela's Orinoco River, and then raised to a productive level.

In the DeLong system, the work deck was a barge hull penetrated by a number of large cylindrical legs or caissons. The legs passed completely through the barge. In their towing position, they projected upward. On site, they would be lowered by jacking them down until each found stable footing on the bottom. Then the barge could be elevated through the use of a pneumatic jacking and holding system. The work area literally lifted itself up the legs to a desired height above the water.

The system had been in use, and it did not require much imagination to apply the same principle to an offshore drilling rig. The seaworthiness of the DeLong design was proven in a 1954 assignment when a DeLong pier was used as a platform to collect core samples needed to help engineer structures for early warning radar installations.

In April 1954, a DeLong-type unit utilizing 12 caisson legs, each 160 feet long and 6 feet in diameter, was put in service as a drilling platform in the Gulf of Mexico. And a larger one followed a year later.

The DeLong design, while basically sound, had its limitations. The multiple legs were massive and costly. And there was the possibility of puncturing a caisson as it set on a rocky bottom. Another issue was wave action in shallow water which could scour support from beneath the large-diameter legs, causing the rig to capsize.

It was left to R.G. LeTourneau, a man who had proven his engineering skills and amassed a fortune manufacturing huge earthmoving machines, to simplify the retractable leg platform.

In 1953, while selling several of his manufacturing plants, LeTourneau agreed to a noncompete clause as part of the deal. Finding himself out of the heavy equipment business, he sought other venues for his knowledge.

Mobile drilling platforms, though far larger than the most massive machine he had ever built, appeared interesting because he believed he had a better solution. His dirt-hauling and logging equipment had relied on independent electric motors at each wheel for power. His system did away with transmissions, conventional brakes, and shaft linkages. The result was a more powerful, rugged, reliable work vehicle.

Using DC electric motors to replace the complex pneumatic leg-jacking and holding mechanism would simplify and improve platform operation.

Further, LeTourneau determined that in place of 10 or more supporting caissons, a properly-sized platform could be braced by as few as three legs. And each of those legs, instead of being a solid tube, could be a triangulated open-truss structure made of a series of welded pipes and bars. Waves would pass through these legs, reducing side stress.

As in all business matters, even the best of ideas had to be sold. It took more than a year of effort to generate an initial order. Zapata Offshore Company, headed by future U.S. President George Bush, was the first customer. Zapata proved to be an excellent client. Its skilled people were able to fit all necessary drilling and materials-handling equipment, along with personnel accommodations, in the available space.

The first LeTourneau-design rig was ready for testing in 1955. The electrically powered rack-and-pinion gears functioned as planned and were capable of handling about eight million pounds. After trials, the rig was towed to Galveston, Texas, where drilling equipment was placed aboard. Then in the spring of 1956 the newly christened *Scorpion* was moved to its first work site. The rig's performance exceeded expectations, and the course of future mobile platform development was set. Both the DeLong and LeTourneau concepts would be refined and improved during the years to come.

Other designs were also tried. Unfortunately, not all were successful.

In 1954 a platform named *Mr. Gus* was the first of its kind. Designed with two separate work areas connected by a walkway, it was intended to operate in up to 100 feet of water. The initial assignment for *Mr. Gus* was to drill at a depth of 50 feet on a location offshore from Freeport, Texas. Problems were encountered while driving steel piles which were to locate and position the structure. The result was a dangerous tilting of the tandem platforms. The unit was brought back into a yard to undergo substantial redesign and modification.

Over the next 24 months, *Mr. Gus* performed without major problems. Then, in March of '57, off the coast of Texas, it capsized.

Another more effective departure from traditional designs was developed by Exxon after years of research. While not truly mobile, it was built on land and then floated on barges to the well site. Called the *Lena*, it is the tallest marine structure ever built.

The concept was a gigantic tower fabricated from large-diameter steel pipe welded into triangulated sections. It was designed to act like a column. With the base resting on the seafloor, the top extended high above the waterline and supported the working areas. To hold the tower upright, guy wires radiated out from the tower structure and were secured to the ocean bottom. In many ways, the idea was similar to the system used to hold up the steel antenna towers of radio and TV stations. Only the *Lena* was built much more strongly.

The Exxon Production Research Company studied the idea during the mid-60s. In 1975 Exxon and a consortium of a dozen other firms constructed and installed a scale model in 300 feet of water off Grand Isle, Louisiana. After 42 months the trial was considered a success, and plans were made to build a full-size version for use at a 1,000-foot depth. The site was to be the Exxon Lena Prospect in the Mississippi Canyon.

Exxon designed and project-managed the effort with Brown & Root as the fabricator and installer. Brown & Root also provided a measure of engineering assistance, working on anchoring the tower base on the seafloor.

Besides the design of the tower, the facility to manufacture the structure and a means of transporting it from the yard to its final location presented challenges in themselves. The sheer size of the tower is staggering. The 27,000-ton tubular-steel monster is longer than several football fields laid end to end. Moving the behemoth from the yard to the barges, which were specially modified to support the structure, was an exceptional feat. It required the innovation of space-age-developed, explosive nut-and-bolt sets to release the tower so it could slip down its skidways into the water all at once. Having one end or the other come off first would have invited damage. Split-second timing was required.

Once in the water, divers ballasted the lower buoyancy tanks and she was towed to the installation site. Another precision exercise was required to sink the device in exactly the right spot and have it facing in just the right direction. In spite of currents and other factors, the *Lena* tower was placed within five feet of the designated mark and was only one degree off its ideal facing.

Setting the 20 guy lines required a highly specialized barge, which alone cost $17 million to create. Each line ran out some 3,000 feet from the tower to a subsea steel-anchor pile that was six feet in diameter and more than 35 yards long. These piles were pounded into the bottom, using an underwater hydraulic hammer.

The completed structure was ready in September 1983. Including decks, derrick, and accommodations, the *Lena* weighed about 47,000 tons. At a height of 1,300 feet, she was some 50 feet taller than the Empire State Building. And a symbol of just how far marine capabilities had come.

One more recent innovation in platform design is worth notice.

The tension-leg rig is used primarily for drilling development wells and is an innovation with its roots in the North Sea. Conoco, in 1981, working with eight other companies, elected to utilize this concept and pioneered its installation in the Hutton field, 90-odd miles northeast of the Shetlands. The selected site was in 485 feet of water, located in a field which was expected to have a relatively short production span when compared with other North Sea positions. So the concept of being able to move the platform at a later date made economic sense, even if the project cost an estimated $1 billion.

Installing a tension-leg platform requires a special template foundation-anchoring system which is set into the ocean floor. Once this is done, the platform is towed into place and correct buoyancy is obtained. Divers secure the platform to the seabed anchor through the use of tubular-steel tethers. The upward pull of the platform, as it tries to float, stretches the tethers, placing them in tension. The taut tethers now serve as legs to locate the platform and maintain its position. The semisubmerged platform gains stability through its design and the tension it holds on the tubular tethers.

The Hutton field platform was brought to operational condition in 22 days after being anchored. This compares favorably with times of nearly a year for more conventional platforms. Conoco demonstrated its satisfaction with the tension-leg concept by announcing plans for a similar unit to be placed in the Gulf of Mexico. Shell has also built a series of tension-leg platforms for Gulf use.

Today the Mobile Offshore Drilling Unit, or MODU, is a sophisticated descendant of original designs. Modern MODUs can be slotted in two classifications. There are the bottom-supported units such as miniplatforms, submersibles, and jackups. Then there are floating structures, or "floaters," which include inland-barge rigs, ship-shaped barges, semisubmersibles, and drillships. In reality, most of the floaters work moored by chains or cables linked to anchors planted in the seafloor.

While inland-barge rigs, the miniplatform with tender, the tension-leg platform, and the posted-barge submersible rigs are still in use, four types of MODUs dominate the current market. These are the jackups, the submersibles, the semisubmersibles, and the column-stabilized units.

The jackup, a bottom-supported type, is the most popular marine rig design in the world. Jackups can be used to drill in waters as deep as 350 to 400 feet and are relatively simple to transport from place to place. The legs can be of the solid-column design if the rig is to be used in waters less than about 250 feet deep. Beyond that depth, the open-truss configuration is needed to offset wave action.

If the seafloor is sufficiently solid, a jackup rig's legs are tipped with spud cans. A spud can is a cylindrical, pointed device which digs into the seabed and establishes a firm grip on the bottom.

In muddy conditions or where the sediment is too soft for spud cans, a "mat" is used. The mat is a strong, flat steel pad shaped somewhat like the letter "A." It is attached to the bottom of a jackup rig's legs. When the legs are lowered, the mat acts like a snowshoe and spreads the load over a wide area for additional stability and sound footing.

The submersible rig, which customarily operates at depths no greater than 175 feet, is a deeper-water offshoot of the posted-barge submersible rig. In place of the barge is a gigantic framework of steel cylinders welded together in the form of a cube. At each corner is a large steel bottle with a tapered top. When the bottles are filled with water, the entire framework sinks to the bottom. When the rig is to be moved, water is pumped from the bottles, making the structure buoyant so it can be towed to another location.

Modified submersibles, with thick metal or reinforced concrete walls around the submerged structure and drilling equipment, are designed for use in arctic waters where ice floes present a serious threat.

The bottle-type submersible gave birth to another type of platform called a semisubmersible, or "semi" for short. Engineers were seeking an answer to drilling in very rough waters and at depths too great for a submersible unit. They began experimenting to discover what would happen if the bottles on a fully submersible rig were filled with only enough water to half-sink the structure. As surmised, the massive weight below the surface delivered a more stable working platform with less tendency to roll or pitch from wave action. Anchors, dug into the sea bottom, were used to hold the rig in position. It was also found that many submersible units could be used as semis.

With the idea tested and proven, marine engineers then designed a column-stabilized unit. Two ship-shaped hulls form the base for this model. The hulls are topped by a number of columns or steel cylinders which extend upward and provide support for the platform. The columns, in turn, are interconnected by crisscrossing smaller steel tubes to add strength. When this structure is fully buoyant, the streamlined hulls allow for easier towing. When the hull and columns are filled with the right amount of water, the rig is partially submerged and becomes stable.

The semi is a well-developed concept which has proven itself in the North Atlantic and the North Sea, where rough waters and storms are the rule rather than the exception.

It is an interesting fact that the underlying designs of the various types of platforms were established and perfected in the Gulf of Mexico. It would require an almost endless number of pages to list all the individuals and companies which contributed to the ability to drill in deeper and deeper Gulf waters. This rough outline of what has been accomplished also does not take into account the human sacrifice and tireless work of those who served on the rigs situated far from land.

While platform advances were being made in the Gulf, work on drillships was progressing exponentially. In place of several contributions from different firms, one California-based company, Global Marine, was responsible for most of the innovating. Its focus was on the development of a highly mobile floater which could drill and work at extreme depths. The company's achievements would provide the means for overcoming some of the most daunting maritime challenges ever encountered.

Chapter 7

Project Mohole: Tripping to Hell and Back

Since ancient times, philosophers, prophets, and poets have placed hell at the center of the earth. Science now assures us a certain kind of hell does exist beneath our feet. And a drilling rig doesn't have to bore very deep to find it. At the bottom of a hole both temperature and pressure build quickly. These conditions were only part of the difficulties to be faced in drilling through the crust of the earth into the area below, called the "mantle."

Earth scientists had spun theories about the makeup of our planet for more than a century before technology made it even remotely possible to "go take a look." One of the early insights in this field of study came in 1909 when an earthquake shook the Kulpa Valley in Croatia. Seismographs across the European continent recorded the event. Many observers noticed a double set of tracings and assumed there had been two closely associated shocks. A Yugoslavian professor at the University of Zagreb arrived at a different interpretation. He postulated that a single shock wave had traveled through the earth at two different rates of speed. One, reverberating in the crust, was slower than the second. This phenomenon seemed to indicate that the earth's outer surface rested or floated on a more dense and flexible layer of rock underneath. His paper, "The Great Earthquake of 1909," contained his calculations of the depth of the crust.

This breakthrough observation, and the explanation, caused a furor among scientists. In time, other tests were made and the results of those experiments confirmed his earlier finding. To recognize the man who first noted the anomaly, the boundary between the earth's crust and mantle was named after him. The "Mohorovicic discontinuity" was a bit hard for Westerners to say or write, so it was abbreviated to "Moho" or merely "M."

In 1956 the American Association for the Advancement of Science weekly journal, *Science*, published a letter from Dr. Frank B. Estabrook, who served with the U.S. Army's Basic Research Branch. He enumerated the scientific value to be found in boring through the earth into the Moho.

The concept presented an enormous challenge because the crust averages almost 125,000 feet, or approximately 24 miles, in thickness. Deep under the ocean, however, the crust is at its thinnest, varying from 20,000 to 30,000 feet. So the problem was one of drilling to a depth of 25,000 feet at a site covered by more than 15,000 feet of water. Quite a task because, at that time, drilling in 200 feet of water was considered a test of technology.

In 1957 the Earth Sciences Division of the National Science Foundation (NSF) began a campaign to expand earth-science research. Money was pouring into the space effort, and this group needed a spectacular undertaking to capture the public's imagination. The members met and reviewed requests for research grants. To their disappointment, none of the 60-plus submitted proposals promised a real thrust to the heart of the most important earth-science questions. Seeking to free their minds to conceive a breakthrough experiment, Professor Harry Hess of Princeton and Professor Walter Munk of the University of California pondered the issue. Dr. Munk then proposed obtaining a sample of the earth's mantle.

To test the audacious idea, Dr. Hess suggested discussing it with members of the American Miscellaneous Society. AMSOC, as the group is called, was created in 1952 out of the Office of Naval Research in La Jolla, California. In many ways AMSOC satirizes more formal and staid scientific groups. The very informality of AMSOC meetings, at which no cow is sacred and discussion of even seemingly frivolous ideas receives serious attention, made an ideal forum for presenting the drill-to-the-mantle notion.

During a breakfast, Dr. Munk advanced his project. Some concepts are offered and languish. Others seem to have an energy of their own and spark human imagination. Among AMSOC scientists, the idea of sinking a hole through to the Moho caught on strongly enough to cause them to form a committee to investigate the matter. A number of distinguished scientists joined the group and a decision was made to conduct a feasibility study. Application was made to the NSF for a grant. The NSF, put off by the unorthodoxy of AMSOC, declined.

Then, in September 1957 at a Toronto meeting, the International Union of Geodesy and Geophysics passed a resolution to study the feasibility of sampling the earth's mantle. During discussion of this topic, a Soviet scientist calmly stated that his nation, having perfected the necessary drilling rig, was at that moment seeking a site.

The Russian's announcement added new vitality to the concept.

A presentation was made to the governing board of the National Academy of Sciences (NAS), which is a respected private organization. Thrilled by being asked to consider a non-space-related experiment of momentous value, the NAS passed an important motion. As a result, the AMSOC Deep Drilling Committee gained prestige by becoming the AMSOC Committee of the National Academy of Sciences.

AMSOC supporters were enthusiastic, and Project Mohole was born.

Drilling to a depth of 25,000 feet by itself presented problems, but had been done. In 1958 a well for Phillips Petroleum Company had been sunk in Pecos County, Texas, to a depth of 25,340 feet. Regrettably, it was a dry hole. The feat did, however, prove such an activity was possible. So half of the Mohole Project requirements could be met.

The unanswered barrier was the water depth. Working in 15,000 feet, or for that matter 1,000 feet, had never before been attempted.

In reporting on what scientific information the Mohole Project might provide, *National Geographic* magazine later noted that samples from the mantle could give insights into the earth's age and makeup. The nature of the earth's magnetic field, issues related to plate tectonics, and causes of quakes as well as volcanic activity were also on the prize list. The publication concluded that a successful Mohole Project would be ranked as one of the century's most outstanding research accomplishments.

There were other values to the program which, when they became known, were not missed by those in the oil industry. Drilling techniques that would be developed and dynamic positioning, or the stationing of a vessel above a selected spot on the bottom without the aid of anchors, could be invaluable in later searches for petroleum.

To make the Mohole Project possible, engineers were charged with finding answers to questions which had never been asked. The drill string, for example, needed to be over seven miles long—five miles longer than what was used in a typical offshore well! No one had any idea of the stresses that would be encountered. And, since a drill string of that length would take several days to come out and go back into the hole, some other means was needed to change drill bits. Reentering the hole after coming out presented an entirely new set of difficulties.

The turning point came when the NAS invited 200 top scientists to a meeting held during a session of the American Geophysical Union in Washington, D.C. Objections to the Mohole idea were raised and discussed. Some felt that information from only one sample from a single site would not be representative of the entire mantle. Others were afraid that such a large expenditure might reduce funds for their projects. But overall there was a sense that, Russian comments aside, it could not be done.

A.J. Field, the engineer employed by Louis N. Waterfall Inc., predecessor company to Global Marine, then took center stage. In place of a lengthy speech, he had a short motion picture. Those in the audience watched enthralled as the *CUSS I* drilled an actual oil well in 200 feet of water. When the film was over, the room was full of converts. For the first time, many realized it might be possible.

With the backing of the National Academy of Sciences

and a number of prestigious professors, the AMSOC committee submitted a new grant request to the NSF. This time, they received $15,000 as a first payment to be used to define the possibility of actually conducting the Mohole experiment.

For the next 18 months, various scientists undertook specific studies of possible sites, drilling technology, and estimates of what knowledge might be gained by the effort. One revelation was that the *CUSS I* was the only drilling vessel in the world that might be modified to do the job. So Global Marine received a contract from the U.S. Navy to define what alterations would have to be made to its vessel in order to carry out a mission of very deepwater seafloor coring. A series of reports followed, and in March 1960, the NSF approved funding for what was to become known informally as Phase I of the Mohole Project. The exploration of "inner space" was on.

In December 1960, a lease agreement with Global Marine for the use of the *CUSS I* was completed. Conversion of the vessel began immediately. In addition to preliminary work on a first-generation heave-compensation system for the drilling platform, which kept the drill string from rising and falling as the barge lifted and dropped with each wave or swell, the company also proposed and improved the dynamic positioning issue.

Shell Oil had operated the *Eureka*, and using a taut-wire line with a sensor on top, had created the world's first dynamically positioned ship.

Global Marine engineers had discussed the problems involved in holding the *CUSS I* at a specified position in waters too deep for anchors, chains, and wire rope. Having seen a New England lobster fisherman use the engine of his skiff to remain stationary in the water while pulling up pots or traps, they got the idea of mounting outboard power units on the barge.

This group connected four 2,000-horsepower engines to a system that linked the throttles with a single joystick control. Moving the stick caused the appropriate engines to reduce or add power, which shifted the vessel in the desired direction.

Position information was attained by using sonar. Taut-wire lines attached to anchors on the seabed held a half dozen buoys eight feet in diameter some 200 feet under the sea. For visual reference points on the surface, four tall spar buoys, which remained in sight regardless of wind and waves, were also tethered in place.

The sonar would ping on the undersea buoys, and results would be displayed on a screen in the control area. In practice, the screen was a bit difficult to read, so another system was also employed.

The joystick operator stood in position and watched the spar buoys. An assistant, following verbal instructions, used a grease pencil to draw a circle on each window around the place where the operator saw a buoy. By nudging the joystick to keep the four buoys in their respective circles, the *CUSS I* was held in position. It was a crude but effective interim solution which could be used while developing a more sophisticated station-keeping technique. A number of other innovations were also added as the Global Marine people engineered answers to questions that arose as work progressed.

Stress on the drill pipe, for instance, was an ever present concern. For this Phase I experiment, the drill string would be over two miles long. To offset any bending tendencies, they devised a special guide shoe. This large, curved, funnel-like gadget extended about 10 feet into the water. The drill string passed through the guide shoe and then down to the bottom where it entered a tapered casing attached to the landing base on the ocean floor. Together, these two devices helped evenly distribute the major bending loads.

By March 1961, the refitted barge had passed all sea trials. On March 7, 1961, the *CUSS I*, crewed by a Global Marine team, was stationed about 20 miles off the coast of La Jolla, California. History was in the making. On March 9 the drill bit reached the ocean floor which lay just over 3,100 feet beneath the surface. Using seawater as a drilling fluid, the crew proceeded to drill five holes. One was more than 1,000 feet in depth. These new frontiersmen had successfully worked in water seven times deeper than the existing record.

A telegram was dispatched to the annual meeting of the

National Research Council in Washington, D.C. Needless to say, the scientific community was electrified. Two of the holes had for the first time yielded samples from the earth's crust which was covered by some 500 feet of subsea sedimentation. And cores with samples dating back more than 15 million years revealed an ocean with far more life than today.

When reports of this stunning achievement circulated in the oil patch, there was disbelief. Then as assurances were received, true elation.

AMSOC's science group was optimistic as well. The operation's first effort had been an unqualified success. They had proven the concept of drilling from a floating, dynamically positioned vessel in water too deep for anchoring. Now it was time to enlarge on their experience with a more rigorous test of the barge's systems.

After a short return to the shipyard for modifications, the *CUSS* was ready. Loaded with special gear, scientists, and interested oil industry executives, she was towed out. For four days, as the *CUSS* was brought to a site near Guadalupe Island some 40 miles off Baja California, Mexico, she rolled and heaved at the end of her towline. With swells approaching 12 feet and winds that reached 25 miles per hour pushing and turning her, the *CUSS* locked into position. Her drilling crew began the long process of making up the drill string and lowering the coring bit, stand by stand, toward the bottom. In more than 11,000 feet of water, the untethered, dynamically positioned vessel held tight to her station. Just after noon on March 28, 1961, the bit turned into the seafloor, starting to sample the earth as it existed millennia before man ever took his first hesitant steps. *Life* magazine later quoted from a scientist's journal of the voyage, saying, "From now on everything is new, everything unknown."

Excitement on board was palpable as the first sample from a hole 110 feet deep was brought up. The clay from 15 to 20 million years in the past was teeming with fossils.

In spite of turbulent weather that tried the vessel's station-holding abilities to the maximum, they decided to drill a second hole. Their goal was to punch into the suspected basalt crust on which the eons of sediment rested. At just over 550 feet into the earth, the drill bit struck a much harder material. In the next two hours they made only five feet of hole. Tension was high as the sample was retrieved. Then before them lay proof of another history-making event. Small pieces of gray, almost black basalt showed that for the first time man had penetrated the hard rock of the earth's crust on the bottom of the sea.

The Global Marine team drilled a total of five core holes while working in waters exceeding 11,000 feet, retrieving almost 30 core samples, including one with 44 feet of basalt. In so doing, they proved dynamic positioning was practical. Using 4½-inch drill pipe and depending upon seawater as their drilling fluid, they defined drilling methods and learned firsthand the challenges that come with working in the middle of a deep, unpredictable ocean.

News of this spectacular achievement had an unanticipated side effect. As the media reported Mohole's successful progress, the project became increasingly famous. Notoriety and seemingly unlimited access to federal funding are serious lures, especially for politicians. So in a short period of time, Mohole became politicized. And a number of influential private companies began pressuring their elected representatives for a share of the action.

A little background is necessary to appreciate what happened after the successful Phase I test drillings.

In 1948 a 40-year-old Texas congressman, Lyndon Baines Johnson, won the state Democratic nomination and became that party's candidate in the senatorial race. His margin of victory over former Governor Coke Stevenson was 80-odd votes, which arrived late in the now-infamous Box 13. Those in the political arena privately said that Johnson had not cheated Stevenson. He had merely outcheated him.

In any case, Johnson went on to win. The U.S. Senate decided not to interfere, and a federal court upheld Johnson's victory.

After losing the presidential nomination to John F. Kennedy in 1960, Johnson surprised many of his close associates by accepting the second position on the ticket. So in 1961, Lyndon Johnson was vice-president of the United States and had spent more than a decade building close support in the Senate. In short, he was a dominant American political figure.

From Johnson's days as a congressman, two supporters, George and Herman Brown, had assisted him both financially and by their friendship. Their company, Brown & Root, was and is a well-respected engineering and construction firm. In addition to building massive dams, laying highways, and constructing naval stations, Houston-based Brown & Root had been active in the offshore oil business since the 1930s. Their expertise ran to piling-supported platforms of various types and underwater pipelines. During World War II, the company also built ships in their yards not far from Houston on Greens Bayou.

The longstanding, well-known personal relationship between the Brown family and Johnson was to play a role in what happened next.

After Mohole's outstanding first-phase accomplishments, the scientific community enthusiastically waited for further progress. Which was unexpectedly long in coming.

Problems had been brewing between AMSOC and the National Science Foundation over which group was to control future work on the project. Friction between these two very different organizations intensified. Finally the National Academy of Sciences, which had backed AMSOC and given that group much needed prestige, found itself caught in the middle of a power struggle. To extricate itself, the NAS asked the NSF to hire an outside, independent, commercial firm to take full charge of future Mohole activity.

This suggestion was intended to shift operational responsibility away from the National Academy of Sciences, which would put that organization in a less exposed position.

In the summer of 1961 that is exactly what transpired.

The NSF, which was, after all, the money source, assumed complete control of the undertaking, leaving AMSOC in an advisory capacity. While the fights for more input and at least some dominion over decisions continued, AMSOC was effectively rendered helpless.

In one of its first acts, the NSF held an informational briefing for any firm that might be interested in submitting a proposal to manage the project. Numerous companies responded. The publicity that continued unabated made Project Mohole exactly what its originators had desired. It was an undertaking that rivaled space exploration for public awareness. In a very real way, Mohole competed directly with the space race as man's crowning scientific undertaking. Firms with special capabilities in matters more down-to-earth than astronautics vied to take part in this new opportunity.

As an aside, the relationship between the two programs became even more publicity-intensive after the NSF assumed control. At one time, there was even a proposal to time both efforts so that the U.S. would land a man on the moon and at the same instant break through the earth's crust into the mantle. Some high government figures considered that such a "double play" would greatly enhance American prestige throughout the world.

Eighty-seven companies appeared at the first Mohole contractor briefing. And many of those later reacted to an NSF request for proposals due September 11, 1961. Global Marine, in a joint venture with Shell Oil and Aerojet General, came forward with a solid proposition.

By all accounts, representatives from Brown & Root did not attend the initial NSF meeting, and the company was not on a list of those which might submit a proposal. Then George Brown became personally interested in the Mohole Project.

By its deadline, the NSF had received a dozen responses, including one from Brown & Root. The top five bidders were selected and asked for more information. Socony-Mobil emerged as the number one contender. Brown & Root was fifth and last on the short list.

For the next step in the selection process, the NSF set appointments with each participating company for face-to-face discussion of its plan. Brown & Root's turn came during January 1962 in Washington, D.C. George Brown took part in the session and apparently made a favorable impression.

Brown & Root did have a background of shallow-water offshore construction experience. This firm was and is a fiscally sound organization with a wealth of technical expertise. And it possessed a proven ability to manage large undertakings as well as maximize the efforts of subcontractors.

On June 20, 1962, the NSF, citing a number of criteria, selected Brown & Root to manage the Mohole job. The total project cost estimate was almost $50 million over a five-year period.

Scientists in the somewhat neglected field of earth research did not miss the fact that this was an unusual opportunity to have their research funded.

Fifty million dollars was enough to make Mohole a sizable government undertaking. And no major amounts of money are spent by any government without political maneuvering to obtain some of the funding for selected constituents. In this case, association with the scientific community added large doses of individual ego. There were blatant attempts to change the entire nature of the endeavor to better suit one or another specialty group. From the start, congressional contention worked to slow or block distribution of the approved dollars.

In spite of all obstacles, the project staggered ahead.

The Brown & Root appointment created a firestorm of protest. Senators lined up on both sides of the issue. Those ardently against Brown & Root argued that the firm did not have adequate scientific strength, expertise in deepwater drilling, or capabilities to do the job. Others supported the selection. The dispute raged hotter and hotter until one member of the Senate demanded that the comptroller general review the Mohole contract procurement procedures. This evaluation concluded the award to Brown & Root was in the public interest. That finding put out the fire, but the unrest remained.

At this same time, the AMSOC and NSF discord had gone public and reached a new high over the best way to proceed on building the drilling vessel. AMSOC advisors were largely in favor of converting an existing, smaller ship, learning from any mistakes, and then building the full-size or "ultimate" vessel armed with that experience. Others had very different opinions. Between the time of announcing the Brown & Root choice and the actual signing of the contract, the AMSOC committee shed its relationship with the National Academy of Sciences and eventually the NSF and Project Mohole. Now AMSOC committee members and Global Marine personnel who had gained a considerable degree of know-how through the Phase I drilling activity were no longer involved in the effort.

To resolve whatever lack of experience they had, Brown & Root organized a separate Mohole division and staffed it with some 50 people. Then they added personnel as work areas were identified.

The site-selection debate was also raging in full force. As time progressed, the contending locales were reduced to two possibilities.

The first was near Antigua, an island in the Lesser Antilles, south of the Puerto Rican Trench. Water there was between 12,000 and 18,000 feet deep. Lack of an adequate on-land operations base and the possibility of being blown away in a hurricane were points against the Caribbean.

The second spot was northeast of Maui, Hawaii, where the crust was thought to be thinner. The island of Maui could provide all necessary support, and only 14,000 to 15,000 feet of water covered the ocean bottom.

In January 1965, the NSF approved the Maui location and set late 1967 as the date to begin drilling.

The challenges were now clear. Working at depths never before attempted, some kind of vessel must maintain position within 500 feet of a given point for possibly as long as three years. And a drill string over six miles in length would have to bore through more than 21,000 feet of very hard rock. Advances in every facet of marine drilling would be required.

Compounding the already formidable difficulties, more and more of what had been private arguments among scientists became public. A large number of researchers who feared funding for their particular projects would be cut back or canceled because of money gushing into Mohole turned to news media to air their concerns. Some complained of the slow progress being made by Brown & Root. Others attacked the program on scientific grounds, claiming that tons of money were being spent for little return. The self-serving refrain was "Why spend for that when you could be funding my efforts which are really important?"

Then there was still the matter of what type drilling vessel would be used. Brown & Root heard the AMSOC arguments for converting a tanker into a drillship. They also considered other designs and settled on a self-propelled semisubmersible platform. The one proven concept, the dynamically positioned drillship, was discarded in favor of an experimental strategy.

Brown & Root's semi design had a pair of lower hulls which were over 30 feet in diameter and almost 400 feet long. The platform was about as large as one and a half football fields. It was estimated to cost $20 million.

To keep the semisubmersible in position during the expected three years that would be required to make the hole, a contract was awarded to the Honeywell Corporation. One Brown & Root engineer, apparently unaware of Global Marine's success with the *CUSS I* during the first phase of Project Mohole, is quoted as saying, "It had never been proven that you could actually stay on location by just having a bunch of engines keeping you on location without any wires."

One by one, other contractors, from major commercial firms to universities to governmental agencies, were brought on board. Which did little to stop members of Congress from bombarding the new Brown & Root Mohole division with an endless stream of applications from still more companies seeking a piece of the money.

Adding to the pressure was the ceaseless probing by oversight groups. Toward the end of the project, monthly progress report meetings lasted three or four days. By the time two weeks were spent developing presentations for the next session, then another week in Washington delivering the information, and yet a fourth week composing minutes of the meeting, there was no time for anything else.

The number of people on the Brown & Root Mohole payroll had grown to more than 200—and that's not counting outside consultants and subcontractors. Attacking each problem systematically and methodically, the team slowly made progress. If a flaw existed in their work, it was the urge to bypass previously developed technology and search for new, promising, improved methods to accomplish the same job. That said, it is also true that innovation after innovation was developed and put to use. From a new way to reenter the hole after withdrawing the drill string to keeping the marine riser line under tension so it would not sag or flex, difficulties were identified and surmounted. Planners even created a new way to drill using a special bit called a "turbocorer."

Finally, in March 1965, nearly a dozen shipyards were asked to bid on the unique semisubmersible platform. National Steel & Shipbuilding Company was the winner, with a proposal which would cost nearly $30 million. Even though National was approximately $15 million under the highest bidder, the price was $10 million above early Brown & Root estimates. Once more, the cost of Mohole had escalated. The original $15 million budget went to $50 million, then $75 million, then $100 million, and finally, in 1966, estimates climbed toward $130 million. There was no end in sight, and not a foot of hole had been drilled. Every dollar spent to that date went to prepare for the final effort. Wags in Congress began calling the project "No Hole," and the press quickly picked up the new name.

To defend the endeavor, the NSF conducted a thorough review. Bolstered by favorable comments from the scientific community, and especially from those concerned with geophysical matters, Brown & Root was allowed to place an order for the semi.

On the other side of the world, however, events were taking place that would contribute to the demise of Project Mohole. Combat in Vietnam had dramatically intensified. Troops and supplies were flooding into the tiny country, and massive construction projects were both planned and underway there. Wars are undoubtedly the most expensive venture a government can undertake. This one was no exception. Members of both houses of Congress were searching for funds, and no expenditure went unquestioned. Mohole was an easy target. And when Congressman Albert Thomas of Texas, who supported the project and served as chairman of the House Independent Offices Appropriation Subcommittee, died in the first months of 1966, a vital ally was lost.

Representative Joe L. Evins, a Tennessee Democrat, assumed the chair vacated by Thomas and began an immediate attack. Dubbing Mohole a scientific mess and an utter waste of capital, Evins moved to kill all further funding. The NSF fought back, supported by a massive lobbying campaign. President Johnson himself used his famous "reason together" approach on several senators, and the project received a brief reprieve. The Senate voted to allocate almost $20 million for 1967. There was great resistance in the House, however. Then the news media struck a telling blow.

Focusing on the longstanding relationship between Johnson and the Brown family, stories began appearing about contributions by the Browns to the Johnson "President's Club." The donations were linked in the press to Brown & Root's Mohole activity, making it appear that the contributions were somehow tied to continuation of Mohole.

Both Johnson and George Brown strongly denied any connection, but the damage was done. Support for the project dwindled. The House did not approve the Mohole appropriation, and when an effort was made to reconcile differences between the House and Senate on this issue failed, all was lost.

In October 1966, the NSF officially ended Project Mohole.

There can be no doubt that the Global Marine successes in Phase I, and the subsequent investigation and development of new drilling and other techniques by Brown & Root, benefitted future offshore oil exploration.

One of the most valuable fallouts from this grand but aborted experiment, however, was human rather than technical in nature. Mohole stands as a monument to the wrong way to conduct scientific research. The NSF departed from its traditional role of initiating and supporting scientific activities. The group assumed the mantle of actively directing and overseeing a program. The resulting infusion of government bureaucracy, patronage, and bickering made a difficult undertaking impossible. And created an administrative disaster.

Chapter 8

A Fleet of Drillships

In 1961, the same year Global Marine's *CUSS I* made offshore history by core drilling successfully in a water depth exceeding 11,000 feet, the company took another giant stride forward.

The market for marine drilling was expanding. And Global Marine was leading the way in deepwater work on the West Coast. Off the California shore, hurricanes were not a threat. Instead, there were earthquakes. Driven by west winds, wave heights could surpass 25 feet, remaining that way for a day, a night, and another day. The coastline was rugged, there were few harbors satisfactory for offshore drilling operations, and onshore facilities were somewhat limited.

Even so, Global Marine found itself with more than sufficient business—and an opportunity to grow larger. To keep pace with increasing demand, the company needed to build additional drillships. And the new vessels had to embody all they had learned from their past efforts. Doing that required capital as well as customers to use the new vessels.

Even though Union Oil was the largest shareholder, Global Marine had successfully managed to receive contracts from Shell, Chevron, Texaco and others. There was some consternation on the part of Global Marine's clients, however, over having a competing major oil company control the firm drilling their exploratory wells. On the other hand, Union Oil had been Global Marine's financial resource.

A partial solution came in mid-1961. Aerojet General

Corporation, a division of General Tire & Rubber active in the missile and space industry, agreed to acquire a 40-percent share of Global Marine. Aerojet General essentially bought half of Union Oil's stock. When the sale was completed, key company managers and employees owned 20 percent of the outstanding shares, and 80 percent was divided equally between Aerojet General and Union Oil. The lessening of Union's position had a positive effect on Global Marine's image among the other major oil companies.

Global Marine management had a vision of creating an entire fleet of second-generation drillships. The updated vessels would begin with the *CUSS II* as a basic design and be built using all the operational experience at the company's disposal.

Bob Bauer; A.J. Field, now executive vice-president and general manager; Russ Thornburg; Curtis Crooke; and their imaginative in-house technical staff went to work developing the innovative features which would be included in the *CUSS II*. A relatively new employee, John Graham, an experienced naval architect who knew little and cared less about the drilling process, had a way with ships. His work would directly influence the designs for many of Global Marine's most noteworthy vessels. The *CUSS II* and the ships to follow it were the very first to be engineered from keel to crown expressly for offshore drilling.

Once the team had the concepts in hand, Graham and his newly hired assistant, Sherman Wetmore, took space in the New Orleans offices of Equitable Equipment Company. Equitable was the firm chosen to oversee construction of what they intended to be the world's largest floating drilling vessel.

Money, however, was still in short supply.

After some effort, a deal was struck with Shell Oil. Shell agreed to a full-payout contract, in the $4-5 million range, to hire the yet-unbuilt *CUSS II*. The amount of that contract more than covered the cost of constructing and operating the ship. So once the *CUSS II* was completed and proven during a series of rigorous sea trials, funds could be borrowed against the Shell agreement. The issue, then, was interim financing—that is, generating enough money to build their new design and perfect it to the point where they could draw against Shell's contract.

Bob Bauer, Global Marine's president, led his management group to Aerojet General to solicit the needed temporary financial backing. In place of money they received disappointing news. Aerojet General's profits—and the company had been very successful—were controlled by General Tire. And General Tire executives took a dim view of investing cash to build an untested drillship.

It is not clear whether this plea for capital instigated the next action on the part of Aerojet General or if a divestiture had already been planned before the request. In any case, during the first quarter of 1962, Bauer, as president of Global Marine Exploration Company, announced the appointment of three new directors to his board. They were Henry L. Hillman, president of Hillman & Sons Company; A. Douglas Hannah, executive vice-president of Hillman & Sons; and Benno C. Schmidt, managing director of J.H. Whitney & Company. Those appointments came about after an intense hunt for funding necessitated by Aerojet General's desire to find buyers who would acquire their 40-percent interest. Henry Hillman agreed to purchase 20 percent and steered Bauer to J.H. Whitney & Company, which took 10 percent. Benno Schmidt, with Whitney, then sent Bauer to meet Allan Shivers, an ex-governor of Texas, at the American Petroleum Institute session in Chicago. Shivers took half of what remained and directed Bauer to two brothers in Corpus Christi, Texas, for the rest. Singer & Singer and White, Weld Company also eventually had investment positions in the firm.

This action was to have a highly positive long-range effect because it was actually the first step toward shifting Global Marine into a publicly owned, publicly traded stock company. The new board members also brought Global Marine management into close contact with individuals in the East Coast financial community. Those associations were later to prove most beneficial.

In the short term, however, Global Marine had a deal with Shell Oil for the new *CUSS II* and a contract with Equitable Equipment Company to build the vessel in its Madisonville, Louisiana, yards. The only missing ingredient was the up-front money to make it all happen.

In those infant days of the offshore drilling industry, financial arrangements were, to say the least, somewhat arbitrary. Deals, counterdeals, and side arrangements were cobbled together as situations arose. Drilling contracts, which naturally originated to handle land-based operations, had clauses changed or stricken entirely to accommodate conditions at sea. The concept of rig downtime is a good example.

On land, there was mechanical downtime, that is, the hours the rig could not operate because of mechanical failures. Compensation for this situation was an industrywide practice. At sea, the mechanical downtime clause was joined by a weather downtime agreement. If weather conditions rendered the rig inoperable, certain discounts to the operating day rate were due. There was no such parallel for land-based operations. Weather downtime rebates, by mutual agreement, were less than mechanical downtime deductions.

Two-tier downtime pricing, naturally, created some intense discussions. If a rig, for example, were inoperable for mechanical reasons and, just as repairs were completed, went down again because of weather, an argument was brewing. The subject would be exactly how much of the total downtime was weather-related and how much due to mechanical ills.

Global Marine developed a type of drilling contract which became the basis for industrywide agreement.

This same financial learn-as-you-go process also made unusual deals and arrangements commonplace, as the development of the *CUSS II* clearly shows.

Global Marine went hunting for the necessary interim money to build its new vessel. Backed by the Shell agreement for use of the rig, management raised the desired short-term capital, and construction on the 5,500-ton, 268-foot-long *CUSS II* began.

The ship they envisioned used diesel electric power and was self-propelled. It carried all the latest drilling equipment. The derrick would be able to handle three, 30-foot lengths of five-inch drill pipe screwed (or "made" as they say in the industry) together into a single 90-foot-long stand called a "thribble." In addition to quarters for 46 people, which could be expanded to accommodate more than 60 persons when necessary, there was a galley and complete amenities for those on board.

An eight-point mooring system, based on a new model diesel-powered winch, would allow operations in waters up to 600 feet deep. In addition, the ship could be swung about when desired, allowing her to meet waves bow-first under most conditions. This changeable placement reduced side-to-side rolling motion. And once on station, the *CUSS II* would be able to make 20,000 feet of hole. All told, she would be the most advanced drillship ever constructed.

Then Shell's people stepped in. They conceived several uses for the new vessel—none as a self-powered ship. Shell wanted a barge which would be towed into position. Needless to say, this preference was not accepted with wild enthusiasm by the Global Marine staff. Shell was the customer, however, so out went plans and equipment for propulsion and in came the necessary towing and other barge fixtures.

Launching of the *CUSS II* in early 1962 almost coincided with a decision by Shell to drill in the Cook Inlet near Anchorage, Alaska.

For that job, a self-propelled vessel was required. So a new deal was negotiated with Global Marine, a company still somewhat strapped for cash.

The *CUSS II* went back on the ways in the Madisonville, Louisiana, shipyard to be converted into a powered ship. And talks began over how such a conversion was to be paid for.

Fortunately, space in the hull had been left during the first change from ship to barge, so powerplant installation was not as difficult as it might have been.

In the end, Shell agreed to furnish and own the necessary diesel engines, the twin 525-horsepower DC electric motors for propulsion, the shafts, props, and rudder. After the Shell contract was completed, Global Marine would be allowed to use those items on jobs for other oil companies and would pay Shell an agreed amount per day until the initial cost was repaid. This was just one more unusual deal made in a formative period.

Launched again, the *CUSS II* underwent sea trials and was able to make some seven knots of speed consistently. With final details attended to in Algiers, across the river from New Orleans, she set out on a 7,146-mile voyage from Louisiana to Alaska. Following passage through the Panama Canal, the *CUSS II* stopped for further outfitting on the West Coast. Then the vessel headed northward, across the Gulf of Alaska to the inlet named for Captain James Cook, an English explorer who first mapped that rugged shore. She arrived in April, after making the trip at an average speed of about 6.5 knots, which equates to a bit over 150 nautical miles per day.

On May 20, 1962, in position with her mooring lines set, the *CUSS II* was ready to begin drilling the Shell-Richfield-Standard Cook Inlet No. 1 well. It was to be a demanding test for a new ship of yet unproven design.

Shell Oil personnel were quoted as saying that Cook Inlet was "the toughest drilling job in the world." In addition to constant water turbulence, tides of up to 30 feet and more come every six hours, and currents run as strong as five or six knots, which was almost the ship's top speed.

Facing these difficult conditions, the *CUSS II* and her crew used all of their resources and managed to drill to a depth of 14,041 feet before suspending operations for the winter.

As cold began pressing in, the ship was brought to warmer, ice-free waters along the California coast. For the next several months, those on board worked with Shell Oil's "Mobot" underwater system.

In the early spring of 1963, the *CUSS II* was again en route to Cook Inlet to resume drilling operations. About 500 miles south of Valdez, in reasonably fair weather, the ship shuddered and shook violently. The engine room crew, fearing a bent shaft or other malfunction, shut down their diesels. Drifting, those on board took stock of the situation. To everyone's surprise, nothing seemed to be wrong with the propulsion system. Then the report came in by radio. A major earthquake had jolted the coast and done serious damage to Anchorage. It had been shock waves, traveling through water, which caused the ship's sudden vibration and banging.

After an intensive inspection which revealed no malfunctions, the *CUSS II* continued into Anchorage, where the quake's aftermath was visible in every direction. Shell Oil management responded immediately, instructing personnel on board the *CUSS II* to assist the city in any way possible. In addition to pipe and water, electrical power was taken from the vessel for use on shore.

The log books on this voyage were later requested by the U.S. government in an effort to obtain an exact time shock waves reached the ship. That information was used by geophysical experts to pinpoint the strength and location of the tremor.

By May of 1963, after the first thaw, the *CUSS II* was back in action. The new well to be drilled was called the SRS Middle Ground Shoal State No. 1. It was again located in the Cook Inlet as a Shell-Richfield-Standard Oil of California joint venture. Working under continuously adverse conditions, the *CUSS II* group brought in the first oil discovery off the coast of Alaska. It was a fitting climax and a tribute to all of those involved.

The advent of new stockholders and a stronger, more diverse board of directors opened several opportunities for Global Marine as a company. Acquiring needed additional capital was given high priority. And realizing management's dream of owning a fleet of drillships was now close to becoming a reality.

As the *CUSS II* sailed on its maiden voyage to New Orleans, Equitable Equipment Company had already begun construction of the *CUSS III* in its Madisonville yard. Work agreements for the new ship were assured, and fabrication proceeded at a brisk pace. By mid-August 1962, she ventured to sea on her inaugural assignment.

The *CUSS III* was a virtual clone of the *CUSS II*, but with refinements and improvements. This time, there was no question of being self-propelled. Cummins diesel engines driving GE generators were installed on schedule. Oceanographic Engineering Corporation of La Jolla, California, supplied the underwater TV capability which could be used for blowout preventer inspection and other tasks. The derrick or "mast" was designed by Global Marine and manufactured by Continental-Emsco.

Pan American International, Pure Oil Company, and Union Oil Company had acquired a gigantic 3,120-square-mile concession off the coast of Trinidad. Working in 180 feet of water, the *CUSS III* was charged with drilling a wildcat well to test the concession. A "wildcat" well is drilled in an unproven field or area without known production. Some believe the phrase comes from the early days of oil prospecting on dry land. An oilman was said to be "out among the wildcats" when he ventured far from civilization and other oil prospectors into the wilderness a long distance from producing wells.

With Pan Am acting as operator for the group, the *CUSS III* arrived at a spot some 20 miles off the island's eastern edge. Seas averaged 10 to 12 feet, and winds remained in the 10- to 15-mile-per-hour range. These were normal conditions for the ship, and no problems were encountered. Working in routine shifts, the drilling crew bored to a depth of 11,580 feet, completed their test on schedule, and returned to U.S. waters. The *CUSS III* then worked off the Louisiana coast for Mobil Oil Company, Humble Oil & Refining Company, and Shell Oil Company. She would later drill off Wilson's Promontory on the southern edge of Australia.

By December 1962, Equitable Equipment's Madisonville yard turned out the *CUSS IV*. Like her predecessors, she motored for final fittings to Algiers on the west bank of the Mississippi. Then, on January 19, 1963, she began her sea career working for Union Oil Company of California in the Gulf of Mexico. Later, in 1964, she would see service in the United Kingdom sector of the North Sea during the summer months and spend the winter season off the coast of Nigeria.

The *CUSS II, III*, and *IV* had been run off as a series. Operating them demanded skilled people who could work every system with confidence. This, in turn meant personnel training, especially for the drilling crews. The new fleet also required a strong marketing program to keep each vessel busy.

John H. Atwood was an ex-professional football player

with an engineering degree from Purdue University. He was brought in to oversee drilling operations and, at the same time, contribute to the marketing effort. During this era at Global Marine, many of the management team wore a marketing hat in addition to the one that went with their specialties. With Atwood came Wendell Williams, another pro ex-footballer, who had gone into sales.

Up to that point, the company's marketing had been predominantly centered on the West Coast. To take full advantage of its fleet, Global Marine would have to strive for business from oil firms operating in the Gulf of Mexico, on the East Coast, and internationally. The success of this effort can be seen by the assignments that stacked up for each ship as it became operational.

The *CUSS V* was in every way a sister ship to the earlier *CUSS II* through *IV* vessels. And she came into being on the same slips and ways as her predecessors. Launched in January 1963, she was ready to perform by the middle of May and set out across the Atlantic Ocean toward the Libyan Gulf of Sirte. *Offshore* magazine reported on both the start of construction of the *CUSS V* and her departure from the town of Algiers on the Mississippi River. Much was made over the ship being self-propelled and highly mobile. And notice was taken of the special restroom facilities that had been installed out of consideration for the religious practices of Libyan national crew members.

The *CUSS V* was constructed because of four 1960 offshore oil concessions granted by the Libyan government. As a result of these concessions, Libyan Atlantic Company had been formed as a joint venture between the Atlantic Refining Company and Phillips Petroleum Company. Libyan Atlantic would serve as operator for the combined 6.7 million offshore acres. The concessions extended as far as 50 miles from land, and water depths were as great as 1,800 feet.

The Libyan Atlantic Company launched its exploration work with one of the largest marine seismic studies ever conducted outside U.S. waters. During a two-year period, more than 5,000 linear miles were traveled during the survey process. Now it was time to drill.

Breaking the waves at her usual seven-plus knots, the *CUSS V*, under an 18-month contract to test the four blocks, arrived and found herself and crew the subject of an intense controversy. The word "cuss" in the Libyan language was a slang word for a part of the female anatomy. Out of respect for the sensibilities of all Arabic-speaking countries, Global Marine coined the word "Glomar" and changed the names of the *CUSS II* through the *CUSS V* to match. (*CUSS II* became *Glomar II*; *CUSS III*, *Glomar III*; and so on.) The by now venerable, and highly venerated *CUSS I*, which would not see duty in Arabic waters, did not undergo a name change.

Another, and by any standard remarkable, difficulty facing Global Marine in those fledgling years was the complete lack of support services. Contracts between the drilling company and the oil company were turnkey. That is, Global Marine was expected to furnish all needed equipment, manpower, supplies, and logistics. Which meant Global Marine had to hire all sea crews, drilling crews, maintenance specialists, mechanics, electricians, caterers, and other manpower. And the company was required to operate anchor boats, crew boats, supply boats, and all ancillary gear.

To help Global Marine personnel acclimate to life in a new country, Taylor Hancock, an attorney and trusted company executive, received many interesting assignments during this period. In addition to his legal duties, he served as the firm's advance man. Before moving staff and equipment into a foreign location for the first time, Taylor would visit each place and compile a fact book. This volume included information on schools, local customs, climate, dress, housing advice, health-care issues, comparative food prices, best restaurants, warnings, encouragements, top bars, spots to avoid, museums, and a host of other practical as well as cultural data. That material was put to good use by Global employees going abroad.

There were no support services because the industry had not yet grown to the point where there was sufficient business to allow firms offering outside resources to be successful. One area which was developing rapidly, however, was the use of helicopters to transport workers to and from rigs. The helipad or landing area for choppers was no longer an optional accessory. It was a vital necessity.

By the early 1960s Global Marine had lived up to its name. The firm was highly respected for its ability to operate internationally. Global was widely regarded as the most advanced deepwater drilling company in the world. It had been a period of great achievement. No one, however, was resting on past accomplishments. The Global Marine team was looking into the future, concentrating on improving all phases of their business and designing a still more advanced generation of drillships.

While they worked, however, events were taking place that would offer challenges which would test the abilities, skills, and ingenuity of every member of this growing company.

Chapter 9

A Time of International
Growth and Innovation

By the mid-1960s Global Marine had earned an outstanding reputation for the ability to drill in places and at depths others said were impossible. It had also proven itself able to operate anywhere on earth. Both of these capabilities set the company apart from competitors and instigated a period of phenomenal expansion.

Only one real barrier stood in the way. The business still needed money—money for new ships and, of growing importance, money for operations. Staffing and supporting its far-flung vessels required a staggering number of employees and unending expenditures for equipment.

To generate the necessary capital, plans were made in 1964 to reorganize the firm as a Delaware corporation. This was the first step in conducting a public offering of stock and debentures. The new name, to simplify matters, would be Global Marine Inc. Going public would also allow continuation of the Union Oil buy out program, which would free the new enterprise from connections with a single oil company. The initial public offering was a solid success and gave the young organization a more substantial position in the business world.

At that time Global Marine also benefitted from another important advantage. Expertise in marine drilling could only be found in the United States. Which meant that anyone wanting to drill an offshore well came to the U.S. to get it done. Since Global Marine had become a standout in its field, the group received a significant share of this worldwide business.

Even with four new drillships working, job offers outpaced Global Marine's service capacity. So an alert management quickly began looking for innovative, affordable ways to increase productivity.

Jackup drilling rigs presented an interesting possibility, and the firm experimented by chartering *Mr. Cap*, a LeTourneau jackup, and having it towed across the Atlantic Ocean. *Mr. Cap* was the first mobile drilling unit to work in the North Sea, off the Netherlands and the German shore. Several months later an operating contract for the *Endeavor*, another LeTourneau-designed platform owned in part by Signal Oil Company, went to Global Marine. It too was towed across the pond into the North Sea.

During this same period in 1964, John H. Atwood, vice-president in charge of Global Marine's Gulf Coast division, hired Robert E. (Bob) Rose. A young man with promise, Rose began his career in operational support. He assisted the *Glomar III* during its Australian assignment and the *Glomar V* in Libya. Additionally he rendered the same services for *Mr. Cap*.

While the jackups performed well, Global Marine's real interest remained with drillships. The *Glomar IV* was contracted to the United Kingdom sector of the North Sea. Rose was selected as an operations manager for this venture. It was a typically massive undertaking which involved transferring work boat crews, helicopter flight teams, housekeeping staffs, rig workers, and all other required personnel to England.

Employees with positions at or above the driller level could serve on "married status" and were allowed to bring their families to reside in England. Those with positions below the driller slot came over on "single status." All lived in the country, with no rotation back to the United States, and worked a 14-day-on, 7-day-off schedule. Which meant there were three complete sets of employees. In all, hundreds of people were involved. So an operations manager, responsible for their welfare, became a father confessor, mayor, police chief, and more to a mixed group of employees.

Conditions were not made easier by the fact that the *Glomar IV*, a highly advanced drillship, was only 268 feet long. Today in the North Sea, there are work boats approaching twice that size. The Global Marine team did not realize that operating from such a small ship in those cold, treacherously rough waters would be as difficult as it proved to be. Nonetheless, they simply did their jobs and met their commitments to Gulf Oil. As a means of avoiding the severe North Sea winters, the ship would be shifted to the Bight of Benin, near Nigeria in West Africa, for several months, then return to the North Sea when the weather was less violent.

The demand for Global Marine services continued unabated, and it was obvious the company could use additional drillships. Based on experience with the *Glomar II* and her sisters and to better meet the demands of harsher marine environments, there was a strong argument for a considerably larger class of vessels. This viewpoint caused serious management debate inside the company. One faction wished to stay with the size ships they were presently operating and build the *Glomar VI*, a clone of the *Glomar V*. A strong argument in favor of that position involved cost.

In the end, a less expensive method of acquiring bigger vessels prevailed. The construction contract for the *Glomar VI* was halted, and the company took one of the most unusual technical directions ever conceived for offshore drilling.

Three older steam-driven ore ships were docked in Beaumont, Texas. They had done time on the Orinoco River and were cheap, if the boilers were usable. Structurally and hydrodynamically, they were well suited for rebuilding into drillships. Their larger size had great appeal as did the engineering challenge presented by harnessing steam to drive a drilling rig.

Hal Stratton's comment that they could engineer it, but no one would be able to run it proved prophetic.

The steam plants on board what became the *Glomar Sirte*, the *Glomar Tasman*, and the *Glomar North Sea*, were relatively advanced closed-loop, marine boiler engines. As such, they were designed for long periods of steady, fixed-throttle, constant speeds.

115

A driller, however, had other requirements. Working at his console, he needed to switch from full power in the drawworks and mud pumps to no power at all in an instant. And then back again to full power seconds later.

Prior to World War II, steam-powered drilling rigs had been in fairly common use—on dry land, of course. The coming of the war, which stopped rig production for the duration, kept many of the old steam units at work. The war's end, however, allowed new diesel alternatives, and even in the early 1950s, steam had been out of favor for several years. Comparatively unclean from an environmental standpoint and with a higher fuel consumption, which contributed to low operating efficiency, steam had seen its day.

Nevertheless, Global Marine engineers took on the challenge. They developed a sophisticated and rather elegant solution to all the problems. Stratton was proven right in the end, though, when maintenance requirements combined with operator-related difficulties rendered the ship's drilling functions more than a bit temperamental. Adding diesels to handle drilling loads and drawworks operations corrected the most serious faults, and the vessels then proved themselves quite serviceable.

From a customer's point of view, however, even with the steam disadvantage, the three new ships were superior. Their larger size, the Global Marine all-chain mooring system which allowed the vessel to be pointed into the waves for greater stability, and increased carrying capacity were all positive benefits. And the ships certainly fit within the money parameters necessitated by Global Marine's financial state. For $18 million and 18 months' work, Global had three drillships that remained leased for many years. After retirement, the *Sirte* and the *Tasman* were converted into tenders. The *North Sea* was sold and finished her life as a drillship. Which demonstrates what a good job had been done with her steam system.

By the beginning of 1966 Global Marine possessed 11 drilling vessels, and 7 of them were less than four years old. Demand for their use had not slowed. In fact, the need for more ships was again apparent. This time, the decision was made to create a program that would allow for the construction of five new craft during the next three years.

To take full advantage of the intense market for their services, speed became a major criterion in building the next class of drillships. John Graham and other engineers started from the basic hull plan of the *Glomar Sirte*. With changes based on past and current experience, the vessel's design was accomplished in a matter of months. Diesel-electric power from Caterpillar engines and GE generators replaced the steam elements of the *Sirte* class. Actual construction of the *Glomar Grand Isle* commenced in mid-1966, and she was delivered in mid-1967. Ninety days later, the *Glomar Conception*, sister to the *Grand Isle*, also came on line. More than 130 feet longer and with nearly twice the displacement of the tried-and-proven *Glomar II* class, the new vessels provided a severely updated definition of the term "drillship." With a multipoint mooring system, which could be used in more than 1,000 feet of water, and a cruising speed exceeding 12 knots, they were simply the best purpose-built vessels of their type on the seas. With few modifications, they were to remain the Global Marine fleet leaders for nearly an entire decade.

On November 29, 1967, just weeks after taking delivery of the *Grand Isle*, Bob Bauer, who in the previous year had been named Global Marine's board chairman, made a very different purchase. He bought the first 100 shares of the new issue of Global Marine stock which was just listed on the New York Stock Exchange (NYSE). With an opening price of $24.63, trading was active, and all involved were pleased. The symbol "GLM" going up on the Big Board signified a gigantic leap forward for the company. In only nine years the start-from-scratch organization had earned both the status and strength needed to become listed on the exchange. Which opened the way for greater access to capital through the financial, as well as investment, communities.

Gaining a spot on the Big Board, though, also brought a predicament. Global Marine, as a corollary activity to its drilling work, had been seeking new applications for the technology it possessed and different challenges for its creative staff.

Global Marine Engineering, which was set up in 1961, was still quite active in the sale of technical services. As a sideline, the company's wealth of marine expertise was, in a limited way, being applied to remote mineral leases and claims that other companies would reject as impossible to develop. Acquiring marginal concessions was not horrendously expensive as there were few other bidders. Global Marine could hold a principal position and solve commercialization problems of a given claim by creating new and effective systems or approaches. The hope was to develop alternative income sources that would bolster company profits during those nearly unavoidable periods when demand for marine drilling was slack.

An opportunity to pursue this alternative venture occurred in 1967. Global Marine was able to gain a petroleum concession which consisted of more than 6.5 million acres of Canadian islands situated above the Arctic Circle. To manage this prospect, a subsidiary company, Global Marine Arctic Limited (GMAL), was formed. In conjunction with Sun Oil Company, Gulf Canada, and other firms, several wells were drilled during the next few years, and GMAL was able to verify the existence of as much as 200 billion cubic feet of natural gas reserves. Economics were not right for further development, and the prospect has to this day not been opened for full production.

Other efforts were made through obtaining small working interests in individual wells in exchange for returning part of the profit realized on drilling services for that job. While none of the early undertakings produced overwhelming success, the premise was sound and management continued seeking opportunities in this arena. Other endeavors included a joint venture to recover phosphate deposits off the coast of Baja California, shrimp farming in Hawaii, and a mine for laterite nickel ore in the Philippine Islands.

The group also acquired gold mining claims in Alaska. Using Global Marine technology, a team was dispatched to dredge and sample the bottoms of several rivers in that relatively new state. As anticipated, traces of gold were found. The activity was not exactly a secret, and those in the company saw no need for

Glomar Beaufort Sea I — Global Marine Development Inc. designed the Concrete Island Drilling System (CIDS) for use in the Beaufort Sea. There are only three ice-free months each year in this treacherous arctic region where temperatures fall to 60 degrees below zero (F.) and it remains dark for weeks on end. The CIDS is completely self-contained and stocks materials and supplies for 12 months of uninterrupted operations.

CUSS I — After purchasing a surplus Navy barge, engineers with Louis N. Waterfall Inc., predecessor company to Global Marine, modified the vessel and from her decks pioneered offshore oil well drilling techniques in use today. The *CUSS I* set numerous records and was truly the "queen."

Glomar Challenger — The *Glomar Challenger* was a purpose-built ship designed by Global Marine to obtain core samples by drilling into the deep ocean floor. Success after success made the *Glomar Challenger* one of the most respected research tools in earth science history.

Hughes Glomar Explorer — The *Hughes Glomar Explorer*, a ship with astonishing capabilities, was built for a top-secret government mission. Her accomplishment is legendary.

Glomar Explorer — The *Glomar Explorer* has now been leased from the Navy and fitted out by Global Marine for her new role as a deepwater drillship.

Global Marine Inc. Goes on the Big Board (November 29, 1967) — In 1967, Robert F. (Bob) Bauer, Global Marine's board chairman, and A. J. Field, Global Marine president and general manager, are shown a tape which registers the first sales of Global Marine common stock on the New York Stock Exchange (NYSE). On the far left is Edward C. Gray, executive vice-president of the NYSE. Far right is Welles Murphey Jr. of Murphey, Marseilles & Smith, assigned specialists in the new issue.

Leadership Evolution — C.R. (Russ) Luigs, right, joined Global Marine in 1977 and served as chairman of the board, president, and chief executive officer. Robert E. (Bob) Rose, on the left, began his career with Global Marine and has now returned to the company as president and chief executive officer.

Glomar Beaufort Sea 1 — Global Marine Development Inc. designed the Concrete Island Drilling System (CIDS) for use in the Beaufort Sea. There are only three ice-free months each year in this treacherous arctic region where temperatures fall to 60 degrees below zero (F.) and it remains dark for weeks on end. The CIDS is completely self-contained and stocks materials and supplies for 12 months of uninterrupted operations.

undue excitement over the limited play. What little gold they found was insufficient to hold serious promise of further development.

Gold, however, is a powerful lure. On the basis of mere rumor, many savvy Wall Street players caught the fever and bought Global Marine stock. In response, GLM's share price started climbing almost immediately after the opening offer. At first, believing the rally was due to investors' faith in the company, management was pleased. Then they learned the cause behind the flurry of interest and realized the situation was less than ideal. As a new issue on the NYSE, it was unseemly to have their stock being traded on the basis of a false rumor.

On December 5, 1967, just six days after being listed, Global Marine management issued a statement in very plain language. In brief, it stated that stories of Global Marine's mining for gold in Alaska were true. The company had received a mineral concession, had done preliminary exploration, but had not discovered a mother lode. In fact, the amount of gold recovered was insignificant. And the claims investigated to date would not produce enough of the precious metal to make any form of mining a feasible commercial venture.

The December 5 denial had virtually no effect on the market. In fact, if anything, disavowal merely added to the furor. If Global Marine had hit abundant stores, so street logic went, the company's management would almost certainly deny it. That would be the only way to gain time to file all legal documents and protect their interests.

In January 1968, less than 50 days after going up on the Big Board, GLM stock was selling for $47, which was almost twice the initial offering price. Executives with the New York Stock Exchange had taken notice of the surge and were closely monitoring the situation.

Despite several attempts to rebut idle gossip with truth, the rush for Global Marine stock continued. Share prices responded. Rumors about the Hunt brothers, noted investors and oilmen, making a big play caused even more interest in the stock. Finally, in June 1968, a high of $63 was recorded.

In response to the seemingly unending speculation, the

New York Stock Exchange suspended all trading in GLM shares for a full day. The same body also mandated a short-term rule against buying those shares on margin. That action stopped speculators from using their credit to acquire Global Marine stock. Margin buying, because investors could purchase large blocks of the stock for smaller sums of money, encouraged rapid turnover and high trading levels. Curtailing margin trading brought some stability to the volatile situation.

Finally, on September 10, 1968, Global Marine declared its Alaskan gold mining venture was not a paying proposition. All exploration had been completely stopped. That announcement had the desired effect, and in a short while, shares were selling for a more reasonable, yet respectable, $30 each.

The desire for an alternative income flow, however, was still present. Global Marine management continued efforts to capitalize on company engineering strengths to gain a working percentage in a variety of mineral interests. To this end, the Global subsidiary specializing in such ventures remained active and, in 1971, was given the name Challenger Oil & Gas.

During the next few years, a staff of geologists and geophysicists was added to provide the company with better expertise in exploration.

As a side effect of growth, entrepreneurial enterprise, and the very nature of the offshore drilling industry, Global Marine personnel were scattered around the world. The resulting decentralization of management began to create operational inefficiencies. To correct this and to maintain the tight company camaraderie which so characterized the organization, Bob Bauer instigated the purchase of a historic 12-story office building in the heart of downtown Los Angeles.

Originally known as the Fine Arts Building, the structure has played an interesting role in California petroleum history. At various times it housed Signal Oil & Gas Company, the Havenstrite interests, Hughes Tool Company, Trico Oil & Gas Company, and many independent producers. In mid-1969 plans were being implemented for Global Marine to move there from its 650 South Grand location and the various satellite sites in the

Los Angeles area.

Renamed Global Marine House, the facility was to serve the firm for well over a dozen years.

As the decade of the 1960s rushed to a close, the company was in an extraordinary position. Financially sound, aggressively creative, and well respected throughout the petroleum industry, Global Marine continued to exceed investors' expectations. While offshore drilling was still the main line of work, two thrilling projects were being undertaken. One was for furthering the understanding and knowledge of the planet we inhabit; the other, for the preservation of our country and our freedoms.

Chapter 10

A Global Scientific Challenge

In mid-1966 the end of the disastrous Mohole Project was drawing near. Members of the National Science Foundation (NSF) involved with the effort had a clear failure on their hands. Frequent criticism from Congress and parts of the scientific community rankled. And the catastrophe was made worse by the continuing and widely publicized success of the U.S. space program. Research money was pouring into the race for the moon. Which meant funding for further earth studies was limited.

A concerned group of scientists had been adamantly against the Mohole concept from its beginning. They strongly believed that a series of cores, taken at various depths from the seafloor at different sites around the world, would provide far more valuable information. When Global Marine's *CUSS I* produced the deepwater cores that had so excited scientific interest, their demands grew louder.

Those dissident scientists who favored wide-scale seabed coring had an excellent argument. The amount of earth knowledge to be gained by such a project was staggering. And for the first time, the technology to obtain excellent samples of the ocean bottom was available.

In 1965, in a test effort to help sell their program, the *Caldrill I* was sent to the east coast of Florida. Working under the direction of the Lamont Geological Observatory of Columbia University, scientists took six cores from the submerged Blake Plateau. One specimen was retrieved from a depth of 3,300 feet. Based partially on results obtained, a unique academic coalition asked the NSF for a grant that would allow an 18-month world-wide activity.

After due deliberation, the NSF agreed and created the Ocean Sediment Coring Program (also called Deep Sea Drilling Project). More than $12 million was allocated for this study.

Four respected institutions were to govern, oversee, and work on the project. These were the Lamont group, Woods Hole Oceanographic Institution, the Institute of Marine Science at the University of Miami, and the Scripps Institution of Oceanography. Together they made up the Joint Oceanographic Institution for Deep Earth Sampling, which came to be known as JOIDES.

Armed with adequate funding, JOIDES principals stated a general goal of gaining knowledge about the formation and history of the earth's oceans and continents. More specifically, they planned to work toward four achievements.

First, they wanted real evidence of seafloor spreading. This would offer a direct indication that the theory of plate tectonics and continental drift was valid. Plate tectonics theory holds that all of the earth's land mass was at one time part of a single huge continent which broke up. The resulting smaller continents we know today are still in motion, floating on the earth's molten core and slowly drifting toward or away from one another.

Second, JOIDES desired cores with fossils that would cast light on the early history of the earth.

Third on their list was the development of serviceable knowledge that could be used in further scientific and commercial marine activities.

And finally, the group wished to demonstrate that drilling and retrieving seabed cores would produce unpredictable but possibly invaluable geological information.

Scripps, chosen by the NSF to manage the project, now faced a serious challenge. This was to be a worldwide effort, so great mobility would be required. That meant creating a special ship which could cruise at speeds fast enough to make changing locations a practical exercise.

She had to have a deepwater drilling capability because some of the sites would be more than 16,000 feet under the sea.

That requirement ruled out the use of anchors, chains, or any other mooring system to hold the vessel in position while coring. The ship had to follow the lead established by the modified *CUSS I* and keep station dynamically, using only computer-controlled propellers to stay put.

Several firms with offshore experience were approached. When it came to marine drilling from a floating rig, however, the staff of Global Marine had the most knowledge. By 1966 this company's fleet included 11 deepwater drilling vessels, and Global was planning to build five more within three years. The *Glomar Grand Isle* was already under construction. Over 100 feet longer and with double the displacement of the previous *Glomar II* class, the *Grand Isle* would have drilling equipment rated for 25,000 feet.

Based on the *Grand Isle* design, Global Marine engineers had been evaluating modifications that would make an extended ocean-drilling mission possible. They were confident they could build and work from such a vessel.

Scripps met with them and received an unusually attractive offer. Global Marine proposed a contract with the NSF to construct a totally new ship. If the vessel performed as guaranteed, Global Marine would be paid. If not, the agreement could be terminated and the company would take a financial hit of some magnitude. To put even more pressure on themselves, they promised the ship in less than a year.

It took 10 months to design, build, and equip the new vessel, which was fittingly named the *Glomar Challenger*. (Her namesake, the British 205-foot *Challenger*, which had crossed the world's oceans using steam and sail power, spent the years between 1872 and 1876 as the first full-time oceanographic research vessel.)

The *Glomar Challenger* was launched on Saturday, March 23, 1968, at the Levingston Shipbuilding yards in Orange, Texas. The wife of Dr. William A. Nierenberg performed the christening, and the ship slid smoothly down the ways into the water. Bob Bauer, Global Marine's board chairman, and Dr. Nierenberg, director of the Scripps Institution of Oceanography, spoke during the ceremony. Those present had high hopes for the vessel's performance and the coring project's future. Few, however, could have even imagined the astounding success the *Challenger* would soon attain.

For five months after launch, this radical ship underwent grueling sea trials before final acceptance by Scripps. With a displacement of 10,500 tons and a length of 400 feet, she was capable of making 12 knots.

The *Glomar Challenger* was a revolutionary design. In a major advance over the *CUSS I*, the Challenger's ability to maintain position above a drill hole in the deepest of waters was a quantum leap forward. Four "tunnel thrusters," huge propellers mounted in "tunnels" that ran the width of the ship, fore and aft, were used in the dynamic positioning maneuver. These units, each powered by a 750-horsepower electric motor, were capable of instantly producing up to 17,000 pounds of thrust to propel the ship sideways.

Sonar signals from a transponder beacon lowered to the seafloor were detected by four hydrophones on the ship. Homing on the beacon, sensors relayed their input to a computer, which in turn operated the thrusters and normal propellers to keep station.

The digitally controlled equipment, provided by Delco, had teething problems. Curtis Crooke, the Global Marine vice-president and engineer on the *Challenger* project, recalls that the system would work well for many hours. Then with no warning, all of the powerful engines would start of their own accord, and the ship would "head for the horizon."

An estimated $500,000 was spent trying to resolve this malfunction. The trouble was finally traced to a glitch on the computer punch tape used in programming. Once corrected, the *Challenger* was able to pass Scripps's severe test. For 72 straight hours, the ship was held in a position which could vary by only five percent of the water depth over the hole. That equates to a variance of only 500 feet when working in 10,000 feet of water. To make the trial even more taxing, no one was allowed to touch any control on the vessel during the 72-hour check period.

As an interesting aside, the *Challenger* also had an older, analog station-keeping system on board. It had originally been designed for the Mohole program but was not used on either project.

In addition, the *Challenger* was the first drillship to utilize a "top drive" system for turning the drill bit, which improved crew safety while adding operational speed. And she was the first commercial vessel to employ the Navy's satellite navigation and communications system.

The *Challenger* was equipped with a derrick that towered 142 feet above the drilling platform on the ship's main deck. It was made to support a drill string that could weigh a million pounds.

The derrick also featured the characteristic shape which had become a Global Marine trademark. The open design of the lower portion of the derrick was necessitated by the automated drill pipe-handling gear, which laid down and picked up the drill string in 90-foot stands from horizontal storage racks.

Along with other refinements, extra motion control was provided by a gyroscopic tank stabilization system below decks.

Finally the *Challenger* utilized the now-proven Global Marine moon pool concept, which allowed the drill string to pass straight downwards from the derrick through an opening in the center of the ship's hull.

Even with all this sophistication, there was some question in the scientists' minds about how the vessel would work as a package. One of the leading experts in the field of stresses on drill pipe felt that pipe life would be very short due to abnormal wear and the length of the drill string. The Global Marine team did not share that view and offered to buy the pipe, then lease it to the NSF-funded management group. That way, the pipe became Global Marine's responsibility. If the pipe failed, Global Marine would replace it. Which would be an expensive proposition, because the ship carried 38,000 feet of five-inch drill pipe made up in thribbles for quicker deployment and recovery. Not surprisingly, the leasing offer was accepted. And the Global Marine team was right. In fact, not one stand was lost due to metal fatigue, and wear was minimal. The knowledge gained in pipe-string dynamics was to prove invaluable to the oil industry and made possible significant advances in offshore drilling.

Global Marine's shipboard crew, which operated the vessel and did the drilling, came to have full confidence in the *Challenger*. Proof that their faith was well placed came quickly.

On August 11, 1968, just a few months after launch, the *Glomar Challenger* hove to far from land in the Gulf of Mexico. The satellite navigational system told the ship's operators this was the place. While hovering in 9,500 feet of water, the sonar transponder was tested by dunking it overboard. Performance assured, the unit was sent to the bottom. Falling through the water, the signalling device took several minutes to reach the seafloor.

Homing in on the beacon, the ship took station and would maintain that position within three percent of the water depth. Station-holding was so good the crew came to take that capability for granted. Months later, during the presidential electoral conventions, a group of off-duty men on board wanted to watch the action on television. But their TV lacked a tuneable antenna. So without losing position, they rotated the ship until the needed alignment was attained to receive a clear TV picture.

Once on station, one of the drilling crews started to send the drill string down. A stand of pipe was made up or screwed into the next, then lowered into the moon pool, where it passed through a conical guide shoe. Rubber collars on the pipe placed about five feet apart helped reduce wear by lessening any rubbing action between the guide shoe and the pipe. Descending into the water 90 feet per stand, the drill string was assembled for hours until it hung beneath the ship two miles or more.

When the bit reached the seabed, contact was measured by sensors on the drill floor. Cables from the drawworks were adjusted to leave just the right amount of weight on the bit to help it bite. Then drilling commenced. And continued daylight or dark. Experience was to show that, depending on depth, between two and five days would be required to complete a core hole.

On this first try, the *Challenger* held station within 60

feet of its desired position. Operations stopped when the drill bit had cut slightly more than 2,500 feet into the seafloor. Cores of subsea layers were taken with a hollow drill bit and a grabbing device, which was sent down the inside the pipe, then retrieved with the core clutched in its fingers.

Those on board were ecstatic. The ship could do the job for which it was designed—and a whole lot more!

On August 19, 1968, from the second hole drilled in 11,720 feet of water, the *Challenger* team pulled up a 472-foot-long core. It was taken from a location known as the Sigsbee Knolls, a submerged string of underwater hills running from near the Sabine River at the Texas-Louisiana border south to the Yucatan in Mexico. Thick, black oil seeped from the sample. And the core itself removed all doubt that the Sigsbee Knolls were in fact salt domes. It was a remarkable find which was to have great commercial impact in the following years.

The first successes were only a small measure of what was to come. Long before the end of the original 18-month term, scientists were unanimously praising the Deep Sea Drilling Project. The NSF, faced with such success, accepted recommendations for more sampling. Regarding the *Challenger's* first voyage as highly productive, they raised their funding to $34.8 million and scheduled another 30-month expedition. In all, the *Challenger* was to carry out her remarkable missions for more than 15 years!

Listing the discoveries made during this decade-and-a-half undertaking would require untold pages of text. Even enumerating the evidence gathered to support major findings is difficult. Here is a short list of the most spectacular results:

Finding:
Scientists were eagerly seeking evidence for the heavily debated theory of continental drift, or plate tectonics. The *Challenger's* first core samples contained fossils dating the seabed at 18 million years. Just 500 miles away, though, the *Challenger* proved that the bottom was 85 million years old. This variance was a clear indication of continental drift.

129

Finding:

Approximately 150 million years ago, Africa and South America broke free of each other. Some 50 million years later, North America, Greenland, and Europe split and, eon by eon, have been gradually drifting farther apart.

Finding:

About 50 million years ago, Australia separated from Antarctica and has been shifting very slowly to the north. How slow? The pace appears to be about two inches per year.

Finding:

Parts of the Pacific Ocean floor once located near the equator 125 million years ago are now just south of the Aleutian Islands, 2,000 miles to the north.

Finding:

During past millennia, the ocean has risen and fallen by as much as 2,000 meters.

Finding:

Over the last two million years, the climate in the Pacific Northwest has never been warmer than it is today.

Finding:

During the past 10 million years, about four inches of the Pacific crust have been consumed each year by the Asiatic continental plate.

Finding:

The Mediterranean Sea is a relatively recent body of water. Only a few million years ago, it was an empty, lifeless desert— a dreary, desiccated depression in the earth some two miles below sea level. Over time, its western wall cracked at Gibraltar, allowing the Atlantic Ocean to gush in and fill the depression.

Finding:

The Gulf of Mexico—offshore Florida, Alabama, Mississippi, Louisiana, and Texas—has gone from a shallow sea of rolling saltwater swells to dry land, then back to a shallow sea again.

Finding:

Severe climatic changes may have caused the extinction of dinosaurs. Slightly over 65 million years ago, plankton may have consumed most of the carbon dioxide-producing plants. Lack of vegetation broke the primitive food chain and led to the disappearance of a vast number of species.

Finding:

The oil discovery on the first leg of the *Challenger's* voyage was only the beginning of locating mineral wealth. Rich deposits of copper, silver, and gold were found in the Red Sea. Precious gems, thrown out by the Amazon River, were noted in the mid-Atlantic. The *Challenger* also confirmed, for the first time, that oil and gas exist beneath very deep waters.

Adding to the *Glomar Challenger's* scientific accomplishments were the practical feats and lessons taught by operating the ship. The *Challenger* was able to stay on station and continue regular drilling activities in 50-knot winds and 15-foot seas. Most anchored vessels cannot match that level of performance in one-tenth the water depth.

The maximum drill string length deployed by the ship, 23,164 feet, was a new experience for all concerned. Nevertheless, it provided information which was extremely helpful to the oil exploration business—as did working in 23,111 feet of water and penetrating 5,712 feet into the seabed on Leg 47. In addition, the *Challenger* managed to drill 3,543 feet into extremely hard basaltic crust on Leg 83 in the Pacific.

Drilling so deep into solid rock from a floating ship required continuous evolution of abilities and techniques. Before the very first coring, Global Marine's engineers recognized that being able to come out of a hole and reenter it was critical. A.J. Field, Global Marine president, likened the task to standing on top of a 10-story building and lowering a piano wire into a pop bottle on the sidewalk. Without reentry, the depth of any one hole was limited to the life of a drill bit because changing the bit meant coming out of the hole.

From the earliest days of the JOIDES project, Global Marine engineers experimented with sonar signals and television cameras to solve the problem. Failures were expected. Failures occurred. But one by one, the difficulties were resolved.

In Atlantic Basin waters 10,000 feet deep at a site some 200 miles from New York City, reentry efforts met with success on June 14, 1970. A special cone with an upper diameter of 16 feet and a height of 14 feet had been landed on the bottom by the drill string. Equipped with three reflectors, the cone became a sonar-visible target. Next, a unique omnidirectional waterjet unit was attached to the end of the drill string. The jet could be sent in any direction with 1,000 pounds of thrust, enough to move a 15,000-foot drill string 800 feet sideways. A rotating sonar transducer, protruding through a hole in the drill bit, pinged on the cone's reflectors. The result was displayed on a TV screen on board the *Challenger*. By watching the display, a skilled operator could "fly" the bit into the cone. With the reentry cone set on the wellhead, replacing the bit in the hole could finally be accomplished. Further work proceeded on this innovative solution, and during Leg 15, the first operational hole reentry was made on December 25, 1970, under 12,982 feet of water. It was a grand Christmas present and, in many ways, signaled the real beginning of ultra deepwater drilling for petroleum.

Sampling techniques improved as well.

During the 96 JOIDES journeys, the *Glomar Challenger* and her crew drilled 1,092 holes at 624 different sites, recovering 19,686 cores. That astounding record is a totally inadequate tribute to the remarkable dedication of those who worked aboard.

Six men who sailed on the first or second voyages were still serving on the *Challenger* for the 96th and last leg on November 20, 1983. Captains J.A. Clarke and J.M. Miles, chief engineer Bert Davis, drilling superintendent A.C. (Junior) Wheeler, bosun Ken Hill, and room steward Reuben Carlos—all had devoted a large portion of their working lives to the *Challenger* and her mission. To scientists, engineers, ocean mariners worldwide, and to the many others who played a role in this amazing quest for knowledge goes a hearty "well done."

Chapter 11

To Go Where No One Had Gone Before, Then Do the Impossible

On a dark, bitterly cold February morning in 1968, not far from the city of Vladivostok, the *K-129*, a Soviet "Golf"-class submarine, slipped silently from its pen. A loud klaxon reverberated through the steel hull, sending the deck crew dashing for their assigned hatches. Once inside, away from the wet chill, they quickly assumed duty stations.

The sub, diesel engines throbbing, sailed directly into the Sea of Japan. Its destination was the far Pacific Ocean. Its destiny would carry it into history.

During the 1960s and beyond, the Soviet Union had the largest submarine fleet in the world. The "G" or Golf-class boats, at least 20 of them on duty in the Russian undersea armada, were first laid down in the shipyards at Komsomolsk and Severodvinsk in 1958. They represented all the Soviets had learned since recovering and studying German U-boats after World War II. Armed with nuclear-tipped torpedoes, the standard G-class also carried three atomic warhead missiles stored in a bulbous "silo" or "fin."

Several of the Golf subs, including the *K-129*, had undergone modifications to improve their firepower. In place of the short range SS-N-4 Sark missiles, these boats were fitted with SS-N-5 Serbs, which had a range greater than 700 miles.

No one is certain of the assignment given to the upgraded *K-129*. We do know its course, direction, and speed. Because like all Soviet craft capable of launching a nuclear strike, long before it came within firing range of U.S. territory, it was enmeshed in an electronic web called "Sea Spider."

In its time, Sea Spider was the most elaborate system ever built for tracking ships both on and beneath the water. Composed of a vast network of satellites, sonar devices, hydrophones, magnetic-field detectors, and a host of other instruments, Sea Spider reached from outer space to the ocean depths. Every vessel, friend or potential foe, that penetrated the boundaries was located, identified, and its position charted on a regular basis.

The *K-129* was picked up in the web shortly after leaving port. During the days that followed, she was monitored on a routine basis. Then, in the latter part of March 1968, catastrophe struck.

Sounds of an underwater explosion, followed by the unnerving screech of plate steel crinkling under tons of pressure, were detected and recorded by SONUS HQ, Sea Spider headquarters in Hawaii. No official explanation of what caused the disaster has ever been offered.

In all likelihood, hydrogen gas, which is produced when batteries are charged, was improperly vented from the sub. Whether the problem was mechanical or due to human error is of little matter. An accumulation of the volatile gas was presumably ignited, and the resulting blast ruptured the vessel's pressure hull.

Those final, tragic moments for the 86 men on board were filled with terror as the boat began sinking rear-end first, deeper and deeper into the black ocean. Their horror was mercifully of short duration. The sub, gathering speed as she dropped, rapidly passed the depth for which she was designed and began to come apart. Death, for every member of the crew, occurred quickly.

Falling, twisting as more metal was torn by increasing water pressure, the boat continued into the abyss. A mile deep, then two, then three, it fell faster and faster. Streamlined for speed through the water, she gained an estimated velocity of nearly 100 miles per hour before slamming into the seafloor. Huge, silent geysers of primordial ooze shot outward for thousands of yards from the point of impact. Then, in this place void of light where time had no meaning, all was quiet.

Monitoring of Soviet naval radio frequencies during the

next few days, as attempts to contact the lost boat went unanswered, indicated the Russians did not know what had happened to the *K-129*. As time passed, their efforts became more desperate. Signal attempts went on a 24-hour, seven-day-a-week, multichannel watch. Fishing trawlers, long known to be equipped with intelligence-gathering electronics and other spy gear, were sent to the lost sub's last known position. From there, they radiated out to cover precisely plotted search grids.

In May 1968, when no trace of the *K-129* was found, the Soviets gradually wound down their activities. There was no choice. The fact that the vessel and her crew had been lost at sea was inescapable.

During this time, the U.S. Navy had not been idle. High-level talks, focused on defining the best course of action, began as soon as it became apparent we knew what had happened and they did not.

By mid-May, the Soviet trawlers were called to other duties. Shortly afterwards, all Russian emergency radio transmissions stopped. The U.S. Navy waited until June to commence a plan which had been decided almost a month earlier. The *USS Halibut*, a research submarine, and the *USS Mizar*, one of the most sophisticated deep-sea reconnaissance vessels ever built, were sent to study the sunken submarine.

With the latest techniques for undersea surveillance, such as acoustical holography, the remains of the *K-129* were located. A detailed survey of the site required almost two months. The broken sub's hulk lay on the bottom in more than 17,000 feet of water. Major portions of her shattered hull jutted upward at a 30-degree angle from a wide, irregular ring of debris.

A complete, highly classified report of the findings was sent directly to the upper echelons of government. And with that report went an absorbing question. What if the United States, acting secretly, were to recover the lost sub?

The salvage idea sparked a series of serious and quarrelsome arguments. Some thought even the notion of a recovery attempt was wild, science-fiction scheming. Others insisted that if it could be done, the wealth of intelligence information would be priceless. The list of possible finds covered Soviet torpedoes, missiles, the warheads themselves, electronics packages, including the internal navigation system, code machines, and a host of other desirables.

The dispute, which involved legal opinions that the taking of another nation's warship breached international law, went clear to the president. Debate lasted for more than a year. Two questions recurred again and again. And without resolving them, the mission would never become a reality.

First, there was the matter of how the submarine was to be retrieved. No one had ever attempted to work at depths three miles beneath the surface. Then, even if it could be done, how would it be possible to keep the act a secret?

Supporters of the mission had an idea about a means to maintain secrecy. And they knew exactly who could be counted on to develop a way to raise the *K-129*.

When they called on Global Marine Inc., the world's leader in offshore drilling technology and engineering, they found a can-do company with a will-do attitude. During January and February of 1970, a contract was let for detailed planning of the operation.

Global Marine moved quickly. Bob Bauer, as chairman, had already taken the matter to his board of directors and gained approval to proceed. A.J. Field, Global president, mobilized the company's best and brightest, including Curtis Crooke, John Evans, John Graham, Sherman Wetmore, and others. They set out to resolve the many obstacles inherent in such an undertaking.

The team worked long hours at top speed without relief. By August 1970, just months after being hired, they submitted a four-volume proposal detailing the design and construction of a system that would perform the recovery. The Global Marine proposition was warmly received and became part of the final presentation used to secure project approval.

The secrecy issue was resolved in an equally expedient manner. If it were impossible to hide an activity, then it was advisable not to try. Instead, it was better to perform the act openly and give the world a plausible story that hid the truth. Then publicize the undertaking until virtually everyone was sick of reading and hearing about it.

The key to their cover was Howard Hughes. One of the world's richest and best known men, Hughes, through his many companies, was already a leading government supplier of top-secret spy satellites and other clandestine electronics. His organization's security clearances were impeccable. And the man himself, a known recluse, held an odd position in the international business community. He was respected for his ability to succeed and make money. At the same time, he was also known as a person who would take chances by using new technologies.

The coverup story was based on truth and, once in place, proved highly effective.

Howard Hughes was creating a minerals company to scour the seafloor, collecting solid nodules of manganese and other nonferrous metals. Existence of the nodules was a proven, scientifically accepted fact.

To accomplish these underwater mining activities, the new firm was building a special, highly sophisticated ship, a massive barge, and a unique undersea-mining machine.

With all major questions answered, the final battle for permission to launch the operation began. Fear of mission failure and a possible confrontation with the Soviets on the high seas was unavoidable. Such an open conflict could lead to a shifting of the delicate balance of power held stable by U.S. policy of detente.

Offsetting the risk was a wealth of valuable information, which could play a profound role in assessments of Soviet military capabilities and future arms limitation talks. If the mission could be done quickly, what was learned might be put to immediate use.

The need for haste did not outweigh the necessity of debate. And debate there was, at every level, continuing into late 1970 and early 1971. Then Richard Nixon, as commander-in-chief, acting on the counsel of his advisors, gave final approval. The Jennifer Project, a joint operation of the Central Intelligence Agency and the U.S. Navy, was a reality. –

Gaining sanction for Jennifer created a flurry of activity.

And established a continuous work marathon at Global Marine. For the next three years, the dedicated group gave up virtually all their family and social lives. Seventy-hour weeks were common, and many pulled 100-hour efforts for months at a time.

The plan called for several components, and Global Marine, while focusing on the ship, also served as prime contractor. In April 1971, construction of the vessel began in the yards owned by Sun Shipbuilding & Dry Dock Company at Chester, Pennsylvania.

At the same time, far to the south in Texas, Hughes Tool was hard at work on forging the necessary lift pipe. Using a type of steel developed for cannon barrels, each 30-foot section was precisely machined and underwent state-of-the-art inspection for flaws. Then every link was stress-tested to loads exceeding 24 million pounds.

Working on a parallel schedule, Lockheed's Ocean Systems Division was producing the massive claw or grapple which would, when lowered to the bottom, lock onto the remnants of the sub and hold each piece fast while being lifted back to the surface. Lockheed also oversaw manufacture of the Hughes Marine Barge, the *HMB-1*, which was being built by National Steel & Shipbuilding Company in San Diego, California. Once the barge was complete and moved to Lockheed's docks in Redwood City, California, components of the complex claw were taken aboard. The entire unit was then assembled away from prying eyes inside the immense floating structure.

As all this proceeded, hundreds of suppliers, from the U.S. and abroad, were being asked to fill special orders on tight deadlines. By this point, thousands of people were involved in some stage of the project. It was time to employ the cover story.

Initial press releases were purposely vague, revealing just enough information to whet a reporter's need to know more. Then Summa Corporation, Hughes's new entity, began to appear in publicity broadsides, ranging from articles in technical journals to stories in business and popular magazines. The theme was consistent. Summa Corporation was a profit-making organization attempting a new and daunting commercial endeavor. Under this cloak, and to swell egos, subcontractors and suppliers were given separate stories about the value of their contributions to the Hughes seabed-mining effort.

Further credence to the cover came from contracts let to firms like American Smelting & Refining to process recovered ore.

As the publicity machine pumped out releases, the story spread nationally, then internationally, until Hughes's venture was discussed throughout the business world.

Other mining companies, being outdone and hammered by the sheer volume of press, came up with plans for their own undersea programs. Giving their concepts to the media meant more ink and airwaves for the revolutionary undertaking. Which in turn lent additional legitimacy to Jennifer's original cover story.

Work on the project's hardware continued at a hectic pace. The Global Marine staff was now part of a new subsidiary corporation created to deal with the government, called Global Marine Mining Inc. Curtis Crooke was named president, and while the corporation cared for legal, economic, and other practical issues, workload stress on the entire team was unabated.

In November 1972, the *Hughes Glomar Explorer* was launched in a half-completed state. Eight months of even more intensive labor were needed to finish the job. In July 1973, she sailed proudly into the Delaware River and under the Wilmington bridge. Then, according to plan, she docked and the top of the high derrick was attached. With the derrick in place, the ship was too tall to cross back under the bridge.

Sea tests began, and as flaws were found, the Global Marine team engineered immediate corrections. Then the *Glomar Explorer*, too sizable for a Panama Canal transit, departed for a long ocean trip around the tip of South America to California.

In the meantime, the lift string, barge, and claw were nearing completion. The *HMB-1* had already performed well in its first trial run near Santa Catalina Island, off the California coast.

Finally, all was ready. The hours, days, weeks, and months of intensive work by the Global Marine staff were now reaching a climax. Laboring against a timetable many outside the company thought was impossible, this group had begun with a clean sheet of paper and devised a plan to do what no one had done before. Then, while overseeing and coordinating manufacture of the hardware, they designed a revolutionary new ship. The *Glomar Explorer* was 619 feet long and 116 feet wide. Employing the dynamic positioning the company had pioneered, which relied upon special electronics and thruster engines, she was able to hold station, hovering above a specific spot on the seabed for days at a time. In place of a keel was a huge opening in her hull, which could be closed by massive steel sliding doors. Measuring 199 feet in length and 74 feet in width, it was an adaptation of the Global Marine moon pool. This opening in the *Explorer's* bottom was the only way to load the immense grapple or claw into the ship. During the recovery process, the lengths of lift pipe would pass through the vessel into the water. And when the claw was pulled back from the dark void with its cargo, the remains of the sub would be brought directly inside the *Glomar Explorer* to avoid unwanted observation.

Only one task remained before sailing.

The *HMB-1* barge, equipped with a retractable roof to conceal the claw that lay inside her, was once more towed to a cove off Santa Catalina Island. Flood gates were opened and she sank to the bottom. Then the *Glomar Explorer* was positioned above the barge, the moon pool doors opened, and the first strings of lift pipe were deployed from the derrick. As this exercise took place, the roof of the barge was rolled back, and divers went down. Working in difficult conditions, they attached the claw to the end of the lift pipe. Once they swam clear, the claw was lifted up into the moon pool and secured there by triangulated supports which ran through from the deck above.

Then, with the grapple safely in place, the moon pool doors were shut and the ship moved off station. When the *Glomar Explorer* was well away and a decent time interval had passed, the barge roof slid closed. Flotation chambers were filled with air, and the *HMB-1* rose again to the surface, where she could be towed back to port.

Finally, after 42 months of grueling toil, all was ready.

One man, John Graham, that talented, taciturn marine architect who had designed many Global Marine ships, including the *Explorer*, and contributed so much to the plan, was not with the team. He had intended to be on the recovery voyage. For many of the punishing months he had been ill. As his cancer progressed, he weakened. His spirit held firm, however, and he stayed with the project regardless of personal well-being. He passed away before knowing the results of his selfless efforts. As a tribute to a gifted colleague, members of the Global Marine team later scattered his ashes over the side of the *Glomar Explorer*.

On July 4, 1974, the *Glomar Explorer* reached its destination in the mid-Pacific Ocean. The ship took station and was ready to perform the task for which she had been built.

One strange twist to this story, however, occurred while the vessel was en route. As matters developed, what seemed a minor event at the time proved to be the first unraveling of the secrecy cloak that had been so cleverly wrapped around the whole mission.

The Securities and Exchange Commission (SEC) had been investigating the Hughes business activities related to a takeover of Hughes Air West. Air West was a small airline flying in the western U.S. and Mexico.

After months of investigation, the SEC was close to obtaining a legal order to force the Hughes interests to deliver a large number of files for SEC review. The Hughes people had strongly contested this action. How the SEC learned of the existence of the documents and where they were stored remains a question. It is probable that some accommodation had been made with a Hughes executive who knew of the handwritten records.

The SEC was successful in obtaining the court order, and it was delivered to the SEC office late in the afternoon of June 4, 1974. The intention was to serve the order the following morning.

On the night of June 4, there was a well-planned and executed robbery of the Hughes communications facility at 7020 Romaine Street in Los Angeles. Published stories indicate that a little after midnight a Hughes security guard stepped outside the building. As he was using his key to reenter through a main door, four men took him prisoner. At gunpoint, he was forced to admit the burglars.

By some reports, only one other person, the night operator at the 24-hour-a-day worldwide switchboard, was in the building. The robbers tied the guard and took him along as they worked
deliberately through the facility. During the next four hours, using pry bars and gas-cutting torches, they methodically opened a
safe and other locked files.

Loss reports later indicated they looted the very papers
sought by the SEC, some other documents, $68,000 in cash, mis-
cellaneous merchandise, and a few salable antiques.

Infuriated by this curious coincidence, the SEC continued trying to serve Howard Hughes himself with papers which
would mandate his appearance in court. This activity is one reason Hughes remained out of this country for the balance of his
life.

The break-in and subsequent events generated a considerable amount of press. There were attempts, or at least reported
attempts, to ransom the papers. To date, however, there has been
no public notice that the files have ever been found. And apparently none of the information supposedly in those files has been
leaked. The documents, like the men who reputedly took them,
simply vanished.

Several responsible individuals at the time expressed skep-
ticism about the burglary. Los Angeles Police Department (LAPD)
officers speculated that the robbers had "inside" help.

One conclusion is unescapable. If the robbery were staged
by Hughes staff members, someone in the Hughes organization
probably had ample warning that the court order had been delivered and knew the time the SEC would try to serve it.

As fascinating as all of this may be, what is the connection with the recovery of the sunken Soviet submarine?

That is the odd twist of fate.

Some days after the burglary, a memo was discovered
missing from the Hughes files. It had not been included in any
list of stolen documents. A search was made without success.
The memo was gone.

That memorandum outlined in some detail the true na-

144

ture of the Jennifer Project and the degree of Hughes's participation in the effort.

Loss of that document set off a chain reaction.

Security of the mission was clearly breached. No one knew if the memo had been somehow passed to the Soviets. The question of terminating the operation must have been discussed. And the political implications of having spent millions to no avail had to have been heavy on many heads. In the end, no stop order was given.

Regrettably, the missing memorandum was the catalyst which blew the cover story apart.

The CIA (Central Intelligence Agency), when notified the memo was missing, began its own investigation. To help with the inquiry, some discussion of the memo's contents took place with high-level LAPD personnel. This, in turn, led to rumors that reached the news media. Rumors were somewhat supported by continuing stories about the break-in. All this set the news hawks to digging. And little by little, more and more was learned. Finally the wall of secrecy crumbled just enough to expose much of the story.

Now for the ironic bit.

Remember the guard? He later admitted taking the memo, hoping to profit by the action. When he read it, the contents so frightened him, he was quoted as saying he flushed it down the toilet.

Imagine the mental state of those aboard the *Hughes Glomar Explorer*. They were alone on the high seas, almost 1,000 miles from anywhere, on a secret mission which might not be secret anymore. Soviet trawlers kept them under continuous watch. Tension on the ship had to have increased as they began to deploy their lift string and the giant claw, reaching further and further into the depths.

One interesting, although unverified, incident occurred when several members of the *Explorer's* crew gathered on deck. They are said to have dropped their trousers in unison and mooned a Russian ship which had come too close to their vessel.

Because the Jennifer Project remains a classified operation, no one can say for certain what was recovered from the *K-129*.

145

The almost slapstick series of events that had unfolded in Los Angeles broke the cover story apart. As the real purpose for the *Glomar Explorer* became known to members of the press, those in charge of keeping the mission secret launched what many today believe was a new misinformation campaign.

Wide news media coverage alluded to the submarine being found in one piece on the bottom, and that all systems on the *Glomar Explorer* deployed as planned. Contact with the sunken boat was made, and the claw gripped its prey. Then, on the way to the surface, one or more "fingers" of the grapple broke, allowing the biggest part of the sub to tear free and fall back into the depths.

This version of the mission raises several questions. Numerous experts tend to think that the three main pieces, into which they believe the G-class sub had broken, were successfully salvaged.

In any case, at least one member of a Senate oversight committee admitted to being briefed by the CIA and was quoted in the press enumerating recovered items. It is, therefore, difficult to dispute that some objects of great value were retrieved—and that the taxpayers received full value for monies spent.

In the midst of this still-lingering controversy, two facts are certain.

First, working against almost impossible deadlines, the Global Marine team redefined our ability to perform deep-ocean tasks. Their innovations and advances opened new vistas for future seabed exploration.

Second, direct economic benefits, which have sprung from work done to make the Jennifer Project a reality, far outweigh the original costs. As an investment, this effort has produced magnificent returns.

One other matter needs touching upon.

There has never been any official recognition of the fact the *Hughes Glomar Explorer* was used for the purpose of recovering a sunken Soviet sub. Also, no official explanation has been offered for why the ship was designed and built. These two comments are important because the *Hughes Glomar Explorer* has, in a very real sense, come home.

Global Marine leased the *Explorer* from the U.S. Navy for 30 years. After suitable conversion, now completed, the one-time mystery vessel which caused so much international curiosity is actively being used as a deep-sea drillship.

In August 1998, just months after her rechristening and working some 175 miles southeast of New Orleans, Louisiana, the *Glomar Explorer* set a new record. The ship and its crew drilled an exploratory well beneath 7,718 feet of Gulf water. Somehow, it is fitting that the *Explorer* has once more begun to conquer the ocean depths.

Chapter 12

From Boom to Bust in the Oil Patch

Every commercial undertaking, at least to some degree, is subject to economic cycles. In the petroleum industry, since the very earliest days, the ups and downs have been cataclysmic. When times are good in the oil patch, they are very, very good. And when times are bad, they are awful.

The last third of this century, beginning in 1970, was one of the most unsettled periods in what may only be classified as a tumultuous trade. Perhaps the best way to gain an appreciation for the highs and lows in the offshore drilling field is to follow the fortunes of Global Marine as the business moved from boom to bust to boom again.

Promising signs in the early 1970s seemed to indicate the coming decade would be a period of solid, yet not spectacular, growth.

Demand for petroleum attained an all-time high with no evidence of a slowdown. Free world oil usage went from 19 million barrels per day in 1960 to more than 44 million in 1972. In that same period, the United States was rapidly using up its surplus capacity, which is defined as the amount of oil that can be produced in excess of actual production. Between 1957 and 1963, U.S. surplus capacity stood at about 4 million barrels per day. By 1970, that figure had dropped to around 1 million barrels per day. The year 1970 also saw American production from domestic wells top 11 million barrels per day. That was a record 12 months. As land-based production began dwindling, however, offshore exploration launched an upswing.

Another factor which would have a pronounced effect on

149

the oil patch came from the rising wave of environmentalism. The 1969 Santa Barbara Channel oil spill was a huge boon to the environmentalist movement. The direct result was limiting oil exploration off the California coast. In addition, that same spill played an indirect role in winning the first of many long construction delays on the trans-Alaskan pipeline, needed to open the Prudhoe Bay fields. And it gave special impetus to passage of legislation mandating environmental-impact statements. This 1970 federal act required an examination of how proposed projects might affect the surrounding natural conditions. The measure effectively prohibited drilling many wells and increased the cost of others.

On the international scene, growing tensions added to the uncertainty of a continued supply of oil from Middle Eastern nations. The resounding defeat of Egypt by Israel in the 1967 Six-Day War, and the aftermath of that short combat, had become a focal point for nationalistic feelings in the Arab states. Renewed talk of oil being used as an economic and social weapon against the West made the world situation uncertain at best. Memories of oil supply problems to European nations during the Suez Canal Crisis in 1956 still lingered. And they were made all the more potent by the 1967 Arab oil embargo against countries friendly to Israel.

So on the evening of August 31, 1969, when a young admirer of Gamal Abdel Nasser, the Egyptian ruler who had seized the canal, staged a coup and assumed leadership of Libya, other governments took notice. Muammar el-Qaddafi's ascendence to power added to the growing Mideast unrest. As would be seen, dependence on traditional supplies of oil from the Arabic nations was to become an unsure course.

The result of these events, and more, resulted in a sellers' market during 1970 and 1971. Which contributed toward even greater interest in offshore deposits.

Global Marine, by this point, was beginning to grow into a complex structure of companies. The core business, marine drilling, was performing well in spite of a market slowdown during the final years of the 1960s.

To a substantial degree, earnings from contracted work on the *Challenger* and the *Glomar Explorer* activities supported company profits. From 1970 to 1974 these efforts delivered about a quarter of the total company revenues.

In 1970 the decision was made to stay ahead of the drillship technology curve, and plans were drawn for a next generation of vessels. When the demand for offshore drilling showed signs of improving during 1971, the company was ready. Construction of the *Glomar Grand Banks* began at the Levingston Shipbuilding Company yards in Orange, Texas.

This new class ship benefitted through lessons learned from building and working aboard the earlier models. When the *Glomar Grand Banks* entered the Global Marine fleet in 1972, she was the first drillship commissioned by the company in five years. Two more of the same class followed the *Grand Banks*. In 1974 the *Glomar Coral Sea* went into service, and in 1975, to fulfill an initial five-year contract with Arco, the *Glomar Java Sea* was commissioned.

It became quite apparent during the early years of the '70s that the United States was no longer the sole source of marine drilling technology. Foreign companies had entered the field and in many cases were doing a credible job. The competitive position of these new entities was enhanced by lower pay scales for those serving in the various crews. Foreign groups also took advantage of more favorable tax treatments as well as added liability protection stemming from registering their vessels under certain flags.

The *Glomar II, III, IV,* and *V*, which had seen heavy duty, were at this point all nearing 10 years of service. While smaller than the other drillships in Global Marine's fleet, these *Glomar II*-class ships were still effective. To make them more desirable and improve their opportunities to gain further contracts, day-rate charges needed to be reduced.

In order to achieve the necessary economies, all four vessels were placed under the Panamanian flag. Crews made up of many nationalities were recruited and trained.

Since Global Marine had previously employed foreign

nationals in many overseas operations, this was not a completely new event. These moves, while giving the company a more competitive position, further internationalized the organization.

Another early 1970's endeavor centered on attempts to develop an alternative income flow by buying into an equity position on an offshore lease. In 1972 the firm took 10 percent ownership on a concession in the North Sea operated by Quintana Overseas Inc. Management also received a one-third interest in a Colombian lease by paying part of the drilling costs through reduced rates for a Global Marine ship. Regrettably, neither prospect proved to have commercially viable deposits. Even in the face of such disappointment, interest in creating a vertically integrated minerals company remained.

Events in the Middle East began to dominate the worldwide oil business during the latter part of 1972. In a long-expected yet still astonishing reversal, the Organization of Petroleum Exporting Countries (OPEC) effectively took over the determination of international oil prices. Then, in 1973, Anwar Sadat, successor to Nasser in Egypt, again attacked Israel, igniting the October or Yom Kippur War. In a matter of weeks, Israeli forces regrouped after the surprise onslaught and, supported by U.S. materials, brought the fighting to a stop on October 26. The cease-fire and subsequent peace talks came not a moment too soon. The U.S. military had gone on nuclear alert in response to potential Union of Soviet Socialist Republics action to lend direct armed support to the Arabs. What began as a limited hostility between Middle Eastern states had quickly escalated into a superpower confrontation more dangerous than the Cuban Missile Crisis.

In a retaliatory move, OPEC members agreed to an immediate oil embargo directed against those nations supportive of Israel, with special reductions in shipments to the United States. Production was cut by five percent immediately and by another five percent per month. In effect, increases in oil prices established by the OPEC ministers prior to the October War were magnified by dwindling world supplies. In weeks, oil went from under $5.50 per barrel to more than $17.00 a barrel.

This direct manipulation of price and supply reverberated through the entire fabric of America, and for that matter, Europe as well. In the U.S., long lines at gasoline pumps spelled the end of high-level customer service, which had been traditional at gas stations. The relief of getting any gasoline at all made self-pumping satisfactory. Drivers were willing to pay almost any price for a tankful. And prices did go up. The age of the service station ended here. From this point on, many became filling stations only.

As the full impact of the OPEC embargo became clear, the U.S. government began a concerted diplomatic response. The goal was to restabilize the Mideast political situation and bring conflicting nations to peaceful discussions. This mediation was effective but took time to work.

Meanwhile the major oil companies had to confront a logistical nightmare. Chaotic market conditions made it almost impossible to plan production of gas, diesel oil, jet fuel, and other commodities in the proper quantities. And with oil costing more, having a surplus of one fuel while users clamored for another would wreck profits. Worse, since trading oil had grown into a fast-paced, rapid-turnover business before the embargo, knowing exactly what was paid for a given barrel of oil in this new economy became a serious challenge. If an oil company missed the price and calculated too low, end products would sell for a loss. If the guesstimate were too high, that same company faced ruinous publicity decrying overpricing and gouging.

To some degree, market forces, including demand and OPEC disagreements, helped balance the situation.

For an average family, higher oil prices meant higher consumer prices. And higher consumer prices resulted in a reduction in oil consumption. That same principle, relating high prices to lower demand, applied to entire societies. If OPEC oil policies became extreme enough to trigger a business depression in the West, demand for their oil would drop. Which would result in reduced revenues. Therefore, causing too serious an economic slump among the mechanized nations would injure OPEC members as well. That reality placed some check on their actions.

Avarice, however, proved to be OPEC's most insidious disruptive force. With international prices so high, an OPEC member could realize fabulous returns by exporting at a rate just a little over its assigned quota. All those involved were to some degree distrustful of one another because of differing individual economic needs, so the alliance became uneasy. In fact, while the official end to the embargo came on March 18, 1974, it had been weakening since late 1973. The full cycle of the embargo's effects on the marine drilling industry, though, was yet to be seen.

The major oil companies at first reacted to soaring oil prices by accelerating their exploration programs to quickly develop new production. Long-term contracts were signed with drilling companies to assure the availability of experienced personnel and equipment.

As the push for new sources of petroleum intensified, the drilling industry responded by adding more capacity to meet what to many, in light of the embargo, seemed long-term needs. In reality, there would soon be a surplus of oil in the marketplace and an overabundance of rigs.

At Global Marine, the true market situation continued to be somewhat obscured by monies being earned through government-funded work. In 1975 company revenues reached a historic peak.

During that same year, the firm was approached by a Norwegian ship company, Einar Rasmussen, which saw a potential in North Sea exploration. Under an arrangement which gave it a small equity position and repaid all costs, Global Marine built, marketed, and operated a completely new semisubmersible rig named the *Poly Glomar Driller*. The venture provided experience in the design, construction, and use of semisubmersibles and was a small financial success.

Going into 1976, the *Glomar Explorer* activity was subsiding and Global Marine expanded the mission of those involved with the project. Earlier, in 1974, the old Global Marine Mining Inc. name had been changed to Global Marine Development Inc. (GMDI). With an engineering staff of between 50 and 100 people depending upon workload, GMDI was a think tank capable of planning and executing virtually any marine engineering project. In a very real sense, this was another move toward corporate diversification and one more effort to generate non-drilling-related income.

Global Marine Development Inc. won several government contracts and, with Curtis Crooke as president of the subsidiary, became an international marine resource.

Among this group's many design achievements was the *Glomar Arion*, an undersea exploration vehicle capable of taking still photos and making high-resolution TV pictures. She was equipped with sophisticated side-scan and avoidance sonar systems.

There was no question, however, that offshore drilling from floating vessels was the firm's main endeavor. So the advent of five-year contracts from both Exxon and Chevron for two new drillships was met with great excitement. This was especially good news because the industry had begun to show a degree of preference for the semisubmersible type of platform rig design.

Even though building the two ships almost simultaneously would place a load on corporate finances, management went ahead. It was a great opportunity to utilize the most recent drilling technology to create another breakthrough ship design.

The fourth-generation *Glomar Pacific* and *Glomar Atlantic* were significantly longer, wider, and had a 1,500-ton greater load capacity than the *Grand Isle* class. With relatively low freeboard, both ships carried thrusters vastly more powerful than those on the *Glomar Explorer*. They were also equipped with an improved, advanced dynamic positioning system. In addition, an eight-point mooring capability allowed bottom-anchored fixed positioning in up to 2,500 feet of water. Each component on the vessels, from the BOP stack to the cranes used for lifting, once again represented the latest and best equipment available.

After sea trials, the *Glomar Pacific* went on duty in mid-1977 and was followed by the *Glomar Atlantic* in 1978.

Another significant event in 1978 was the retirement of the *CUSS I*, the drilling barge that initiated marine drilling. The *CUSS I* had reigned as queen of the Global Marine fleet. Over and above being the experimental vessel from which basic floating-platform offshore drilling was developed, the queen and her crews scored first after first. Her performance in the testing phase of the Mohole Project, where she'd held station dynamically and obtained core samples from unprecedented depths, was only one accomplishment. In late 1965 the *CUSS I* had established another record by drilling, running, and cementing multiple strings of casing in 632 feet of water without the aid of divers. Though small by current standards, the old barge was a match for larger vessels. From Alaska's Cook Inlet to California, she worked the West Coast and did jobs for most of the major oil companies.

During one assignment, the state of Alaska offered a bonus by reducing royalty costs for the operator whose rig was fastest at penetrating commercial oil sands. The *CUSS I* was arrayed against the *Glomar II* and a vessel from another company. Limited to the use of 60-foot stands of pipe called doubles while the others used 90-foot lengths, the *CUSS I* still managed to earn the bonus by being first.

In the aftermath of the Santa Barbara oil spill and the subsequent halting of all offshore activity in California, the *CUSS I* sat idle from August 1969 until she was refitted for more use. On September 20, 1974, she began drilling a well for Standard Oil of California. She remained active until her final retirement March 22, 1978. While no exact statistics are available, many believe the *CUSS I* drilled more offshore hole than any other marine rig. Ever.

After decommissioning, she was sold, and there was talk she would resume life again as a seagoing fish cannery.

Some vessels manage to create a place in the hearts of those who serve on them. So it was with the *CUSS I*. As she sat quietly at Pier D in Long Beach that final March day, a member of her crew, no one is sure who, left a note tacked to her galley bulletin board. It read, in part, "To the men who worked on her, she was the Queen. We were proud to work on her. She may have been small, but she always did her job." The note went on to state, "She was such a lady, sometimes moody, as ladies will be, but she always did her job. There's a lot of sweat and blood on her decks and patches on her hull, but to us, she is still the Queen."

The retirement of the *CUSS I* was noted by several oil patch trade magazines with equal solemnity and thanks from a grateful industry.

By late 1976 and into '77, the worldwide oil market was still seeking equilibrium amid ongoing upheavals caused by the OPEC embargo and price setting. The OPEC nations themselves were doing little to bring price stability into a volatile commodity business.

Dwindling revenues from government contracts, coupled with a distinct slowing of demand for offshore drilling, had a pronounced negative effect on Global Marine as a company. In 1976, for the first time in corporate history, and then again in 1977, Global Marine ended the year with a financial loss.

Faced with $150 million in debt, cost overruns on construction of both the *Glomar Pacific* and the *Glomar Atlantic* and a fleet in which 10 of the 16 ships were over a decade old, the firm was headed for a difficult period of adjustment.

To make matters worse, foreign competition in the drilling field was greater than ever, oil companies from other nations were growing in scale of operations, and the semisubmersible rig design was gaining even more market preference.

As might be expected after two consecutive years of losses, there were questions from the company's creditors. It was time for Global Marine to reexamine its market position, analyze its goals, and reposition itself for future expansion.

That difficult process was approached with the same "let's get it done" attitude which has characterized all Global Marine activities. A key element was the transition in administrative direction from an entrepreneurial style of leadership to one based more on the skills of professional managers.

In May 1977, C.R. (Russ) Luigs came aboard as president and chief executive officer. Luigs, bright, knowledgeable, and articulate, was a sturdily built man with a self-deprecating, friendly manner. A degreed petroleum engineer, he had begun his career in oil tool sales with a company in East Texas. He also spent five years as a sales engineer in Venezuela. By 1974 he had become president and chief operating officer for U.S. Industries,

157

which owned a large number of subsidiary companies. As part of his duties, Luigs had helped those firms shift into professional management modes, so he was well equipped to contribute to Global Marine's internal realignment.

After months of intense effort aimed at fixing what was wrong and keeping what was right, a revised course was set for the company. Global Marine now defined its primary service as offshore drilling, not offshore drilling from drillships. The drillship was still an indispensable tool, but the new definition clearly showed that working from other platforms would not be a sideline proposition. To back that position, a program was established to diversify the fleet by adding semisubmersibles and jackups. Immediate actions included a revision of banking alliances; the sale of the *CUSS I*, the *Glomar II*, as well as the *Glomar IV*; and the conversion of the *Glomar Sirte* and the *Glomar Tasman* into tenders.

Perhaps the most sweeping resolution of this period, however, was the decision to increase drastically both the size of the company and the number of vessels available for contracts. To accomplish this expansion, new financial resources had to be found. Locating the required money was not an easy task. Global Marine was already carrying a debt-to-equity ratio that exceeded three to one, and total company equity was less than $45 million. Losing money during the previous two years made the hunt for capital more difficult. But it had to be found if Global Marine were to maintain its leadership position in the industry.

Internally, the realignment produced Global Marine Inc. as the parent company and five subsidiaries, each a separate profit center with its own goals, operating budget, and president.

Global Marine Drilling Company was the largest of the five and would function on a day-rate basis, charging customers for time spent working in their behalf.

Global Marine Development Inc. (GMDI), headquartered in Newport Beach, California, continued with government commitments, marine engineering projects, and offshore construction contracts.

Oceanographic Services Inc. monitored and evaluated

arctic sea icing and weather conditions along with other ocean conditions in support of high-latitude operations.

Challenger Minerals Inc., the new name for Global Marine's exploration and development arm, continued to pursue opportunities in various mineral prospects.

The fifth subsidiary, Applied Drilling Technology Inc. (ADTI), was formed in 1979 and demonstrated that Global Marine had lost none of its respected creativity. The goal was to deliver turnkey offshore drilling services. A "turnkey" job promised an oil company a completed, logged well, drilled to an agreed depth at a fixed price. This mode of payment was common in land-based drilling contracts. In marine drilling, though, because of the added downtime variables from weather, it was most unusual. Meticulous planning was required, which necessitated a highly experienced staff. A single omission, a miscalculation, or a defective piece of equipment could result in huge losses on a project.

ADTI was to prove quite successful. The turnkey marine drilling concept gained acceptance, and the company was a pioneering offshore provider of this unique service.

In the past, Global Marine's efforts to diversify had waxed and waned in concert with the ups and downs of the drilling market. When drilling was off, attention was paid to the subsidiaries. When demand for drilling was high, focus shifted from alternative revenue sources back to the main area of operations.

The revised corporate structure alleviated this problem. Each subsidiary operated under the Global Marine umbrella as an almost autonomous, independent company.

By 1978 the new strategy was being implemented. In business, as in so many other endeavors, timing is vastly important. Those changes came about at a most propitious moment.

By late 1978 it was clear the drilling market was about to make an enormous upturn. Global Marine's resolution to grow larger and improve overall performance meshed almost perfectly with this singular opportunity.

Capital for expansion remained a problem, however. In early 1978, aided by the fact that the books once again showed a profit, the company began to attract additional financial resources. And new lines of credit came through countries which wished to support their national shipbuilding capabilities. Demand for offshore rigs of all types translated into an appealing alternative for the yards, and rig fabrication was considered highly desirable business. So governments including Canada, Singapore, France, Finland, and Hong Kong offered financing to Global Marine at attractive interest rates. The U.S. Maritime Administration was also involved and acted as guarantor on some notes.

As part of its expansion and diversification program, Global Marine purchased a relatively new Norwegian semisubmersible. Rechristened the *Glomar Semi I*, she commenced drilling operations off the East Coast shortly after the sale was completed. Funds were also attained to begin construction of the LeTourneau-designed *Glomar Jackup I* in a Canadian yard. That rig was to be at work by the middle of 1979. Other monies allowed building and placing two rigs on Chevron's Platform Grace in the Santa Barbara Channel. By late '79 those rigs were engaged for 24 months to drill 48 wells.

As the 1980s came closer, the exploration boom was expanding at a prodigious rate. Demand for hydrocarbons drove the need for new offshore discoveries and greater production from existing fields. Global Marine's fortunes rose with market needs, and the company began ordering construction of several jackups at a time. Costs were contained by the use of off-the-shelf plans.

The diversified, burgeoning fleet brought on a rash of name troubles. In an effort to simplify rig identity and provide a better description of the size and operational scope, a new procedure was enacted. Only a limited number of designs were employed to keep modifications, maintenance, and repairs as easy as possible. Among the jackups, for example, all new rigs were either LeTourneau 82SDC, LeTourneau 116C, or Friede-Goldman L780 designs. These three types became the *High Island*, the *Adriatic*, and the *Main Pass* class rigs in the Global Marine fleet. The first *High Island*-class rig acquired by the company received the name the *Glomar High Island I*; the second the *Glomar High Island II*; and so on. The name told the rig capability; the number generally revealed the order of coming into the fleet.

Semisubmersibles were classed as the *Biscay* or the *Arctic*. "Biscay" referred to the Ocean Victory design and "Arctic" to the Pacesetter line from Friede-Goldman. All names were preceded by the familiar word "Glomar." So the *Glomar Adriatic II* was the second Marathon-LeTourneau 116C-design rig acquired by the company.

To say that business was good during this period is a gross understatement. Desire for offshore services of all types made price secondary to equipment availability. Shortages of rigs and capable crews pushed day rates higher and higher.

In support of the main line of business, the plan which realigned management of the subsidiary companies began to pay off. In March 1979, Windsor Producing Company sent Global Marine a check for $7,525.03. The amount was not as important as the fact that it was the first money earned by Challenger Minerals Inc. from owned production. In addition to wells in the Giddings Field, Lee County, Texas, which produced the revenue, the company also had a five percent interest in new gas reserves recently discovered in the Canadian Arctic, gas wells in Central and West Texas, and three offshore wells in the Gulf of Mexico.

Global Marine Development Inc. was also busy. Its engineering expertise had taken it far afield from the oil business. One project centered on the *Glomar Explorer*, the massive mystery ship with very special capabilities. The vessel's originally stated purpose had been to recover metals from the seafloor. And the ship was well designed for that task. Refitted with a prototype system developed for Ocean Minerals Inc., a Global Marine team took the *Explorer* out to sea again—this time to accomplish her cover-story mission.

The effort proved the feasibility of recovering commercial quantities of minerals from the ocean bed. But an international agreement furthered by Third World countries prohibited staking bottom-of-the-sea claims. So the entire issue became moot. Without the enforceable legal protection of a claim or concession, no company or nation wished to spend millions to make a find, then watch others reap the benefits.

Another significant GMDI activity included the development of a unique floating, liquified natural gas (LNG) production and storage facility. Made from concrete, which reacted well to the extremely cold LNG temperatures, this movable factory could be transported to an offshore field where it would process natural gas, compress the gas until it liquified, then store the finished LNG product in the utmost safety. Sales efforts were made to the Iranians, Indonesians, and Mexicans. Regrettably, there were no takers.

Government contracts continued to be another focus of GMDI activity. Some were classified. Others involved work with alternative forms of energy. One experiment centered on a floating facility to grow, then harvest and process seaweed into natural gas. There were also an offshore windmill farm and numerous other ocean-related projects.

One of the largest federal contracts deserves special notice. In yet another display of technical brilliance, GMDI's team accepted the challenge of building an ocean thermal energy conversion (OTEC) plant.

An old U.S. Navy tanker was moored directly over a lava seabed off the coast of Hawaii in 4,500 feet of water. To hold the ship in place, a ball of used anchor chain weighing almost 250 tons was sunk to the bottom and attached by a steel cable to a buoy on the surface. The ship then used the buoy as its mooring point.

A boiler and all necessary piping were installed on the ship, then filled with a liquid which would boil at about 80 degrees, the temperature of the surface ocean water. Seawater heated the boiler which boiled the liquid, turning it into vapor. The vapor was piped through a turbine, turning it to generate electricity. Once through the turbine, the vapor passed into a condenser where it was chilled by colder water sucked up from the ocean depths. Lowering the vapor temperature returned the gas to liquid form, and the process was repeated.

The system was, in essence, another means of converting solar energy, which warmed the upper layers of the sea, into electricity. To work, the procedure required approximately 100,000 gallons of chilly seabed water per minute. To handle this enormous flow rate, three plastic pipes, each four feet in diameter, hung down from the ship and reached more than 2,000 feet into the ocean. The system proved workable but too costly to operate. So unless energy prices dramatically increase, OTEC will not be a commercially viable source of renewable energy.

Global Marine continued building its fleet to support its main area of business. In 1979 the company ordered four Marathon-LeTourneau 82SDC jackups at a total cost of $110 million. Three of those rigs were to be built by Marathon Manufacturing Company in Brownsville, Texas, for delivery in 1981. The fourth unit was scheduled for 1980 completion in the Davie shipyard of Quebec, Canada. Davie Shipbuilding Ltd. was already at work on two other Global Marine jackups of the same design.

To help pay for these capital outlays, Global Marine, with 4.4 million common stock shares outstanding, filed an SEC registration statement to issue an additional 700,000 shares through Drexel Burnham Lambert Inc. Coincidentally, five days after that filing, on May 30, 1979, a Challenger Minerals prospect paid off. A gas well, located in Block A-567 in the High Island area, in which the firm held a 42.3 percent working interest, came in with 108 feet of net gas zone.

The year 1979 also saw Global Marine stockholders receive their first cash dividend. Announced at the company's annual meeting, shareholders of record on July 16 received $0.05 per share. Also of importance was the solicitation of shareholder approval to split the common stock two for one, bringing the total to 12 million shares.

By any measure—be it profits, revenues, stock price, or equity—Global Marine was on a roll. Between 1977 and 1982 earnings soared from a $6 million loss to $85 million in profits. Assets rose to more than $1.3 billion, and cash from operations exceeded $200 million. During the five years, sales increased 500 percent! Regrettably debt, while well in control, had also grown from $160 million in '77 to approximately $680 million in '82. Even so, all the company could do was strive to keep pace with demand. Experienced, qualified people were woefully difficult to find, so manning the rigs demanded constant training of newer personnel. And virtually all necessities, from mud to drill pipe to bits, were in short supply.

A strange, almost insane state of euphoria and optimism infected those working in the oil industry. Even knowledgeable experts were carried away by the intensity of the good times, and several notables predicted that the boom would last for decades.

In Houston, Texas, the offshore oil capital of the world, companies were doing business at a fevered pitch. The annual Offshore Technology Conference, known as the OTC, is a week-long meeting of those involved in marine operations around the globe. It went from a rather staid exhibition of oil-related equipment and the presentation of technical papers to an extravaganza. In this booming period, some firms budgeted close to a million dollars for their sales display and hospitality rooms. In 1980 over 80,000 people attended! Firms were flush, the folks working for those firms were flush, and cities where oilmen gathered prospered as well.

It is difficult to communicate the almost arrogant confidence that seemed to infect every level of the industry. Banks were actively encouraging their customers engaged in the drilling business to borrow more money and build more rigs. Offshore and on, the entire drilling world appeared to be exploding. It was as if those in the oil companies controlling exploration had gone temporarily mad, along with the banks which had once been so conservative. On land and at sea, the Cajun French expression, *laissez les bons temps rouler*, let the good times roll, became the catch phrase of the Gulf Coast.

On May 27, 1980, Global Marine stock split two for one. The company also announced its third straight year of record revenues, and since a quarterly dividend of $0.05 per share was paid after the split, the dividend was effectively doubled.

The stock split was a broadside fired as part of an ambitious plan by management to grow the Global Marine fleet to a total of 52 rigs by 1985. That feat would involve adding two drillships, 25 jackups, and seven semis, for an increase of 34 new units in five short years. And the program was well underway.

The price tag for this expansion was expected to exceed $2 billion, plus another $700,000 for acquiring oil and gas properties through Challenger Minerals.

Obviously cash, and lots of it, was required. As mentioned earlier, part of the funding support was to come from foreign governments assisting their shipbuilding yards. The stock market, along with company credit, earnings, and other resources, would provide the rest. A bit over $6.5 million was gained through the sale of the Global Marine House building at 811 W. 7th Street in Los Angeles.

The as yet unannounced goal of the challenging program devised by Global Marine president, Russ Luigs; William R. Thomas, senior vice-president for finance and administration; and other key Global Marine managers was to build sales from $200-plus million in 1980 to more than $1 billion by the end of 1985.

A number of national magazines, alerted to Global Marine's growth strategy, wrote favorable feature articles. The company's stock remained strong. In November 1981, a *Wall Street Journal* story suggested caution in buying shares of land-based drilling firms because of a possible rig oversupply. The warning specifically did not apply to the offshore companies.

Wall Street players, however, did take notice when in October of '81 Global Marine paid an effective 20 percent in long-term interest to place $80 million worth of 20-year bonds. Note was also made of the problems that existed in obtaining loan guarantees from the U.S. Maritime Administration for work in U.S. shipyards. Even so, the company posted record earnings in quarter after quarter.

Seeking still better performance, Global Marine Development Inc. was charged with directing its technology assets toward breakthroughs which would make substantial contributions to the parent corporation's profitability. The GMDI team decided to pursue three possibilities: arctic engineering, ultra deepwater production techniques, and shallow water production systems.

Many interesting and innovative concepts were developed, tried, and improved. An air-cushion icebreaker vehicle, which was tested and easily broke through sea ice three-feet thick, was reported in May 1982. Use of this unique device, which could extend the Beaufort Sea drilling season from three to five months, was offered to the industry. Earlier, in 1980, GMDI had applied the air-cushion concept as the basis for an environmentally sound new type of drilling platform that floated on air above the water surface. These designs, along with an air-cushion arctic utility vehicle, were only three cold-climate concepts to come from this prolific group. They also drew plans for a deepwater production tower for use in depths between 2,000 and 10,000 feet.

One true innovation, though, did result in added corporate income. Alaska's Beaufort Sea had been open for drilling since the mid-1960s and contained proven reserves. Offers of new leases caused excitement among the major oil companies.

The environment there, just south of the polar ice cap, is a driller's nightmare and presents serious dangers. The ocean can freeze from the surface to a depth of five or more inches during a single night. In the 60-below degree Fahrenheit temperatures, ice can build to a thickness of seven feet. Winds drive surface ice into jams more than 10 yards high, and these floating mountains crush all in their path. In the upper latitudes, winter is marked by sunless darkness broken by curtains of northern lights. All inhabited areas on the rig must be enclosed from the cold, and supplies need to be stocked against the serious possibility of being trapped by ice for several months.

Standard practice was to explore by drilling from an artificial island built with dredged gravel or a temporary island made of ice. If the potential find merited the huge expense, a larger gravel island would then be constructed as a base to develop the field. One of the costliest nonproducing wells in history was drilled in the Beaufort Sea by SOHIO. So certain of the exploration data on its Mukluk prospect, SOHIO bypassed exploratory drilling and built a huge gravel island in anticipation of producing its predicted find. Costs for this failed attempt were said to be in excess of $1 billion.

Prior to the SOHIO misfortune, a GMDI team, including Sherman Wetmore, who had primary responsibility for arctic engineering development, created a portable concrete structure which could be towed to a site, remain in place as long as needed, then be moved to another location. Called the Concrete Island Drilling System (CIDS), it offered drastically lower costs and other advantages over the traditional practice.

The CIDS is an integrated assembly which includes a steel mud mat, a patented honeycomb of reinforced concrete to deal with ice, and a pair of steel deck barges. At 95 feet tall and 300 feet on a side, the unit can be sunk to rest on the sea bottom and may be stocked to contain all necessary supplies to drill and care for those on board for up to a full year.

This concept was proposed to several of the major oil companies. Although some expressed interest, none placed any orders.

Then GMDI struck pay dirt. Exxon had encountered problems that interfered with plans to establish its island. In addition to difficulties in building roads to allow construction equipment access to its site, there were environmental concerns over the use of gravel in the completed structure.

On the basis of a handshake, Global Marine Development set out to make its CIDS a reality. Exxon needed the unit in a little over a year, so it could be sealifted and installed in the desired location during the limited Beaufort Sea summer work window. That left fewer than 12 months to build the revolutionary concrete island.

Financing for this undertaking was rather complex. After much negotiation, a private investment group owned the concrete "brick," a land drilling contractor owned the drilling equipment, and Global Marine leased both the brick and the drilling equipment and held the contract to staff and operate the unusual vessel.

GMDI turned to Nippon KoKan in Tokyo, which accepted the challenge to build this prototype arctic exploration rig in record time. Work began in five different yards simultaneously.

The arrangement with Exxon was a full payout contract which would cover all costs. So provided the rig worked as promised, the financial risk was limited. There was one other major worry in the agreement. If Global Marine did not deliver the completed CIDS by the July deadline for the sealift, the company would have to relocate the unit to a safe harbor for storage until next year. That meant no income for 12 months to offset debt payments on a device which cost in excess of $100 million.

The CIDS, now christened the *Glomar Beaufort Sea I*, was successfully completed in 1984, which allowed on-time delivery to Exxon. Over the next decade the rig performed flawlessly, drilling two wells for Exxon and one for Arco in ice conditions unmanageable for almost all of the world's 550 other offshore drilling rigs.

By early 1982 the industry's frantic pace began to slow inexorably. A limited economic recession which had been striking areas of America on an erratic basis spread to most of the nation, resulting in reduced energy demand. Diminished use was particularly felt in the natural gas arena, where prices were miserably less than expectations. Oil prices then began easing off, causing both major and independent oil companies to cut back exploration activities in response to tightening cash flows.

Throughout the offshore industry, and indeed across the entire oil patch, managers took stock of their positions, realizing there was a glut of rigs of all types. Easy credit, an insatiable market which cried for greater drilling capacity immediately if not sooner, and the wellspring of profits resulted in still more offshore units in various stages of construction around the world.

As the year progressed, signs of a drilling downturn became more foreboding. In response, Global Marine stepped up its sales efforts to acquire additional long-term contracts for its rigs.

Company activities were somewhat hampered at this point by a planned shifting of the firm's headquarters from California, where Global Marine had been born, to Houston, Texas. Houston had become an international oil center and was the heartland for worldwide offshore activities. The move placed Global closer to vital clients and suppliers.

Fortunately the stepped-up marketing activity had a positive effect. A peak of nearly $1 billion in drilling backlog contracts representing future revenues provided what appeared even to conservative observers as a highly protected position. To add further support, the company began stockpiling cash, building reserves to nearly $350 million, and at the same time had access to $120 million more from lines of credit. Global Marine faced the uncertain future with a firm financial base. Industry experts predicted that 1985 would see the end of the down cycle and the beginning of a new uptrend. Throughout the oil patch, the motto became, "Stay alive until '85."

What no one suspected was the unprecedented tenacity of the decline that was then already in progress. There had been slow periods in the past. While not enjoyable, careful planning and operating economies had made them bearable for most solid firms. This time, though, the vicious severity and length of the slump was unparalleled. Before it was over, the entire drilling and oil service industry would be in chaotic collapse.

The coming of 1983 saw the company further aligning itself to weather hard times. Without slacking off its sales activity, Global Marine management reviewed all rig construction then underway. Several orders were canceled and others, where practical, had delivery dates extended as far into the future as possible. All that could be done at that time was done. Still, there was one inescapable fact. The yards were delivering more and more rigs into a market which a member of the Global Marine board of directors described as already having "just one rig too many." Despite its efforts, the company accepted seven new rigs that year.

Adding to management concerns was the fate of the *Glomar Java Sea*. Under contract to Arco, she was the first American marine drilling contractor's ship to work in waters claimed by the People's Republic of China.

While moored on location in the South China Sea drilling their third well, those aboard the vessel had followed the weather reports relating to Typhoon Lex. Charting the storm's course made it clear they were in for a very tough blow.

According to a contemporaneous newspaper report, the ship communicated with Global Marine's Houston operations center about 10:10 A.M. Houston time on October 25, 1983. It was 11:10 in the evening Peking time. The captain's final report noted 75-knot winds and a 15-degree list to one side as the crew, who had donned life jackets, fought turbulent seas. Then all contact with the beleaguered vessel was disrupted by the storm. One more message was received in Houston. It was an SOS. Not long after, Typhoon Lex passed directly over the drilling site.

As an immediate rescue mission was organized, a second distress signal was picked up by a Japanese merchant vessel the next day. Then only silence. As preparations were being made, hopes for the safety of the crew rose and fell. A Chinese tug in the area found no trace of the *Java Sea*. Some extra buoys marked the spot where she had been moored, which led to the conclusion the captain had ordered the vessel freed of its mooring lines and had made a run through the storm. The same Chinese tug located an oil slick and four life vests.

The news gave management several chilling possibilities. The *Java Sea* might have entered Vietnamese waters and been captured. Or since the Vietnamese disputed China's rights to the area, they could have attacked the ship. Then there was always the chance that an old mine might have struck the vessel. Added to those worries was the fact that many observers noted the *Java Sea* had an appearance similar to the *Glomar Explorer*. What if a Soviet sub had spotted her and mistakenly assumed it was the *Explorer* out on another foray. Or equally as bad, knew the *Java Sea* was a Global Marine ship and sought revenge for the role the *Explorer* had played on its clandestine mission? From a position under the towering waves, a sub would have been safe and stable even during a typhoon. Was it possible that a Soviet submarine had used the storm for cover and torpedoed the *Java Sea*?

The notion could not be ignored and added another dimension to the need for understanding what had taken place.

Global Marine, assisted by several other international firms and agencies, set out to find and rescue any survivors—as well as learn the cause of the disaster and honor those who might have been lost.

After weeks of intensive work, including the use of divers and an expenditure of millions of dollars, the results were clear. The ship lay on the bottom, cracked almost in half, with a fracture running from the bilge keel to the main deck. Samples of recovered steel from the vessel and extensive visual analysis revealed no sign of torpedo attack. She had foundered in heavy seas and gone down with all hands.

In all, 81 souls were lost on that bleak October day. Of that number, 42 were Americans and 35 were Chinese nationals. According to newspaper stories, 60 of the crew were employed by Global Marine, and the rest of the personnel worked in specialist capacities for various companies, as Arco, Schlumberger, Sub Sea International, The Analysts, Halliburton, and Dresser Industries.

This sad event affected the hearts of the entire Global Marine family. Both past and present Global Marine employees continue to offer condolences to the relations and friends of those who passed away in this ocean disaster.

For the balance of that traumatic year, Global Marine's rig utilization remained high, at over 90 percent of the fleet's capacity, despite industrywide problems. Gross revenues were down for 1983, 1984, and 1985, but only slightly. Reevaluations and discounts on the value of Challenger Minerals' oil and gas properties, however, had a negative effect on profits.

The real fiscal threat came from debt service. With fleet growth stymied and rig utilization as well as rates locked into a declining mode by slackening demand, revenues, while temporarily stable, would deteriorate. The amount owed on long-term borrowing, though, was fixed and exceeded $1 billion.

Profits in 1982 remained substantial with the company earning $85 million. In 1983 earnings fell to $49 million. In 1984, after the industry had been in the doldrums for two years, there was a loss of $91 million. Then, in 1985, with no end to the drilling downturn, the loss reached $220 million.

Global Marine responded, staging a series of belt-tightening moves to conserve cash through lower operating expenses. In addition to stopping dividends, it became apparent, as contracts were fulfilled, that rigs would have to be placed in storage, or "cold stacked" as the practice was called. This process was delayed as long as possible. If the hard-cash cost of stacking a rig was $2,000 a day and the operating expense for the company to run the rig was $10,000 per day, Global Marine could accept an $8,000 per day contract, which would result in keeping its crew employed. Under such an arrangement, the company suffered no greater cash loss. When even that became impossible, as the price of oil continued to slump and drilling was reduced by 50 percent, rigs, including some with working contracts, were stacked around the world.

International events were about to make a return from the brink of financial ruin an impossibility.

Saudi Arabia had taken a unique position among the OPEC nations. By altering their production level of oil, the Saudis were able to defend the OPEC-mandated price per barrel of crude. The country, however, was paying for its role as leader. In the early part of the 1980s Saudi Arabia harvested almost $120 billion from oil exports. By 1985 it received only a bit over $25 billion. And its output of oil fell below what was taken from the U.K. sector of the North Sea.

The drop in income meant canceling or at least delaying ambitious programs of public works that were underway. Worse, to Saudi leaders, their loss of income diminished their country's power on the global stage and reduced their prestige among the Arab states.

The other OPEC participants took advantage of Saudi Arabia's adherence to agreed production levels and began ignoring the quotas to boost their own treasuries. Saudi ministers went from asking for cooperation to action in a matter of months.

Their plan was ingenious. They offered the major oil companies a new pricing deal. Saudi Arabia would guarantee refiners a profit of $2 per barrel and take their payment for the crude oil from part of the earnings made by selling the end products to the consumers.

The program caused an OPEC uproar. And this one action almost resulted in the complete collapse of the worldwide drilling and oil field services industries. Keen observers realized that the cessation of Saudi Arabian support of oil prices would cause a drop in the cost per barrel. Most, though, believed the decline would be within reason. Most were wrong.

One grade of U.S. oil from West Texas was selling for $31.75 in mid-November 1985. In the first quarter of 1986, the same oil was going for about $10! Some oil from the Persian Gulf was available for under $6. Predictions of $5 oil were met with concerned plausibility.

It required time for the OPEC states to regroup, and their efforts to stabilize oil prices were hindered by the winds of war blowing between Iran and Iraq. Within a year, they began to reassert control over pricing, but it was an uneasy, delicate balance. And the damage had already been done.

At Global Marine, personnel cuts, office consolidation, disbanding of subsidiary companies, rig stacking, and other economies were not enough. With revenues faltering, there was no possibility of cutting costs sufficiently to meet financial obligations. In 1985 debt service alone was running approximately $240 million annually. And by late that year, it had become apparent the company must have some form of debt restructuring because cash reserves, while substantive at about $90 million, were not undepletable.

Contract backlog, including short-term, low-margin deals, had dwindled to around $150 million, and signs were not promising for rebuilding this amount.

In July 1985, Global Marine ceased debt service payments. After serious attempts to find a means to restructure its obligations met with no success, there were only two avenues left. One was total liquidation, which was not palatable. The debt level was about twice the value of company collateral. And there was some question, in view of market conditions, about being able to sell parts of the fleet at all. The other possibility was to seek protection from creditors through the court.

On January 27, 1986, Global Marine filed under Chapter 11 of the U.S. Bankruptcy Code.

The company was not alone in that perilous time. By 1986 the domino effect was toppling one business after another. No drilling meant no use for drilling rigs, which meant no use for drilling fluid or other rig equipment. As those markets died, so did all support and ancillary services from supply boat operators to helicopter companies to catering firms.

Houston, as a city, was hit desperately hard. House repossessions became commonplace. Whole subdivisions which had been home to hundreds of families suddenly contained rows of empty dwellings facing streets where sheets of newspaper blew in the wind. And there were no buyers for those properties. Companies collapsed, leaving behind tens of thousands of square feet of unoccupied office space. Unemployed executives with little prospect of finding work in their fields of expertise no longer had expense accounts or health insurance. So restaurants shuttered and the medical profession was made aware of the pinch.

Many banks and financial institutions, including some of the largest in the South and Southwest, had double trouble. In addition to an excess of failed loans to oil-related customers, real estate values were falling, and defaulting on property notes was commonplace. So one by one, they too began to collapse. By 1986 malaise in the drilling industry had spread to impact every economic segment of the community, and beyond. From Brownsville to Baton Rouge, the oil industry-driven Gulf Coast was sunk in an economic depression whose echoes reached far into the next decade. The new slogan became *lache pas la patate*, Cajun slang for "never give up."

As soon as the filing was complete, Global Marine management immediately began working on a plan to bring them out of Chapter 11. Protection helped, but drains on company cash continued. Revenues for 1986 fell to $186 million, about half the 1985 figure. The net loss for '86 showed a little improvement, at $209 million. One bright spot was settling a lawsuit centering on a take-or-pay natural gas contract, which brought in approximately $80 million.

For 1987 revenues dropped by more than $100 million, to $82 million, and the loss was $132 million, which cut deeper into cash reserves. By May 1987, 27 of the 30-rig fleet were still idle. There was some improvement, however, in the latter part of the year when 17 rigs were employed. Day rates were down, though, to just pennies over direct-cash operating costs. And the end was still not in sight.

Revenues for 1988 were up, but the net loss stayed at the 1987 level and cash was again reduced. Still, progress was being made with all concerned in the Chapter 11 proceedings. Lenders took note that the company's cash position was holding and the fleet was diversified as well as relatively new. Most creditors were also well aware of market conditions which had hounded Global Marine through the past few years and realized the oil services industry had virtually collapsed.

Through careful planning, fair negotiations, and reasoning with all involved, company leadership was able to offer a restructuring plan which won approval.

On February 28, 1989, Global Marine Inc. emerged from Chapter 11 protection. Battered by the brutal realities of the catastrophic world oil markets, yet still standing while many competitors had folded, management looked toward better times.

Chapter 13

Into the Next Cycle — Genesis

On February 28, 1989, after difficult months of work, Global Marine emerged from Chapter 11 protection. Its fleet was virtually intact. So the company had the most up-to-date, well-maintained equipment available from any firm in the marine drilling industry. Its crews were among the best. It had top-flight, in-place, operating and financial controls. The organization was leaner, meaner, and the employees had their pride, mixed with the expected Global Marine can-do attitude.

It also found itself in a very different world.

The drilling slump that became noticeable in late 1982 proved to be unrelenting. The oil field service business vacillated between poor and awful for nearly seven straight years. And in that time closings, mergers, acquisitions, and bankruptcies had changed the face of the industry. Many of the oldest, most respected names were gone. The remaining firms were, on the whole, larger in size and smarter in terms of utilizing their resources. The competition for what few jobs were available was fierce.

The highly mobile Global Marine fleet had been strategically dispersed about the world. Rigs were located off the North Slope of Alaska, in the Gulf of Mexico, in the North Sea, riding the waters near Africa, off Australia, and in the Mediterranean. Most were working although day rates barely covered operating costs.

Top management had developed a business plan which built on their previous efforts. The task was to generate a com-

177

plete recapitalization, including further adjustment to company debt. This program would result in a financial position which would allow the company to better deal with the cyclic nature of the business.

In spite of limited capital, the technical people had begun equipping the fleet with an improved drilling system which relied on top drives. The first unit was installed for testing in 1987. A top drive, as the name implies, turns or drives the drill string from an electric power source suspended high in the derrick above the drill pipe. Increased safety and improved performance are only two of its benefits. While no longer used as a power source, the rotary table is retained as a means of supporting the drill string in the hole. The switch to top drive was a strong demonstration of Global Marine's intention to maintain its position of leadership.

In 1991 the company added a heavy-weather jackup rig and employed it in the North Sea, where it could work at depths up to 300 feet.

The most important occurrence in 1991, though, was the implementation of the Global Marine Incentive Drilling Program.

Developed in 1989 and then tested during 1990, this innovative concept was a harbinger of the full quality effort the company would soon undertake.

Incentive drilling was a means of lowering the cost of a well to a client while gaining a bonus based on how much the client saved. Rates for rigs in the Gulf of Mexico hovered around the $45,000-per-day level. Those in the more severe North Sea were costing the client about $125,000 per day. In addition, the client's "spread cost" for all other necessary services was two to three times more than the day rate. So any proposal with promise of reducing the length of time required to drill a well was welcome.

In Global Marine's Incentive Drilling Program (IDP), the customer's well plan, which called for making so much depth each working day, was carefully evaluated. Then company goals, including safety, were set. If Global Marine beat the figure and drilled to the ordered depth before the planned number of days expired, the client paid only for the days actually spent drilling—plus a bonus which was to be divided between the crew and the company.

Shortly after formally introducing IDP, six programs were in place and 15 proposals were being considered.

In spite of mediocre market conditions, 1991 was an active year for Global Marine.

The *Glomar Arctic I* went into the yards to receive top drive and added mud-pump capacity as well as other modifications. When completed, she moved on assignment to Amerada Hess Corporations's Scott Field for a long-term contract.

Global Marine, through a new entity, Global Marine Oil & Gas Co. (GMOG), was also exploring on its own. GMOG contracted with its fellow subsidiary, Applied Drilling Technology, to provide turnkey wells in the Main Pass Block 69 Prospect, three miles off the Louisiana coast. That effort was moderately successful.

The company finished 1991 with a profit, but there was little cheering in the hallways. Too many indicators pointed toward a softening market for 1992.

In a consolidation move, Challenger Minerals and Applied Drilling Technology were placed under a single president and began operations from the same premises. Challenger was now responsible for managing the existing assets it held.

As the year progressed, one bright possibility became a reality. In 1986 Global Marine had filed a legal action against Transcontinental Gas Pipe Line Corporation (Transco), a subsidiary of Transco Energy Company, over a take-or-pay gas contract. Transco had contracted to take 100 percent of Challenger Mineral's gas from High Island Block A-567 or pay if they did not take it, at prices which turned out to be well above the market. Global Marine delivered the gas as required by the contract. Gas prices had slipped further into the tank, however, and the spot price was only $1.50 per million cubic feet. When the client refused to pay the required contract price of over $7.00 and negotiations proved futile, Global Marine went to the courthouse.

Nearly seven years later, on May 8, 1992, the lawsuit was finally settled. Global Marine received $20 million in cash, another $20 million in the form of a note to be paid over eight installments, and $15 million in shares of Transco stock. Needless to say, the money was welcome and improved Global Marine's financial position.

At the close of 1992, incentive drilling contracts had provided well over $1.5 million in revenues to the company and employees. There were 27 vessels in the fleet. And work was underway to improve internal communications through the development of a wide area network communications system linking Houston with the North Sea; the Aberdeen, Scotland, office; the facilities in Lafayette, Louisiana; the Gulf Coast Materials Center; and Applied Drilling Technology.

The major turn in the company's fortunes, however, was made on December 23, 1992. Global Marine provided details on a successful restructuring of the company's debt and finance. C. Russell Luigs, CEO and chairman of the board, stated, "Completion of the recapitalization marks the culmination of a 15-year effort to position Global Marine as a preeminent offshore drilling contractor." He added that the company had a first-class fleet, a first-class organization, and now a financial structure compatible with the cyclic nature of the industry.

On the oil business side, natural gas prices had strengthened and shown signs of improving still more. The impact of this positive news was almost immediate. During part of 1992, activity in the Gulf of Mexico had slowed to levels not seen since 1986. Contracts were so sparse that Global Marine was forced to stack three of its five rigs stationed there, leaving only the *Main Pass IV* and the *Adriatic II* working. Traditionally business increased during the fourth quarter. That year, expectations were for only a mild improvement.

Then nature struck. There is an old saying to the effect that an ill wind blows no good. Hurricane Andrew, sweeping across Florida, devastating all in its path, slammed into the warm fall Gulf waters. More than a hundred platforms were damaged just as the gas market ignited. As a result, increased demand for drilling rigs seemed to come overnight.

Global Marine responded by unstacking the *High Island* units *II* and *IV*, then still needed more. For the first time in ages, demand and day-rate economics made it advantageous to move rigs into the long-blighted Gulf of Mexico.

Adding to the excitement, a Dallas, Texas-based company purchased the *Glomar Biscay I* and began an immediate conversion for locating it in Garden Banks Block 388. As part of the transaction, Global Marine signed a 10-year contract to manage the refurbished rig, which left Malta on January 14 and arrived in Pascagoula, Mississippi, March 8.

To add further capacity, the *Adriatic IV* was dispatched from Italy to the Gulf, and in October went to work for Enron. Demand remained high, and because of the rig shortage, day rates became especially favorable.

While the Gulf was starting to heat up, the count of rigs in use throughout the rest of the world was falling. In response, Global Marine ordered the *High Island III* and the *High Island VIII* from West Africa and the *Main Pass I* from the North Sea. Each arrived safely and in time to meet contract obligations.

Suddenly the size of Global Marine's Gulf of Mexico operations made it difficult to believe that only a year or so earlier three out of five rigs had been idle. Personnel assigned to the Gulf increased from about 150 to an eventual 725. To keep ahead of possible staffing difficulties, the company placed added emphasis on its training program to maintain service and safety standards.

Then, on August 4, 1993, in the midst of the American offshore explosion, management turned to Wall Street. In all, 17.25 million shares were offered and taken. The sale gave the company $67 million, which was used to retire the $25 million mortgage on the *Glomar Baltic I*, expand the fleet, and fund a joint venture. This was the final step in debt reduction, cutting the figure from $1 billion to $225 million in seven and a half years. That 80-percent decrease slashed debt service to $29 million annually from a high of $240 million per year in 1986. With the resurgence of activity in the Gulf, a strengthened balance sheet, and cash to spend, there were deals to be made.

Seventeen million in cash and the *Moray Firth I*, a severe-environment North Sea jackup, were exchanged for three Marathon-LeTourneau 116C jackups. The three newly acquired rigs were rated to work in up to 300 feet of water and had the drilling platform cantilevered over the stern to gain greater utilization of on-board space.

Not long after, in December, another $30 million bought two more of the same class rigs. This resulted in the acquisition of a total of four new units for under a third of their replacement cost. The Global Marine fleet now boasted 23 premium jackups, all rated to drill in 250 or more feet of water.

As Gulf of Mexico activity continued to increase, demand for rigs slackened in other areas. Utilizing their quick-response mobility, Global Marine drew in seven more of its jackups. Five came from the North Sea and two from West Africa. All arrived in good shape, and shortly thereafter, the company had 12 premium rigs operating off the U.S. coast.

To some degree, increased exploration drilling in the Gulf of Mexico was spurred by the arrival of a new technology. Entrepreneurs had come back to the oil patch. This time, though, they brought with them software instead of the steel machines and hardware of their predecessors.

Through the years, the techniques to perform seismic surveys, which reveal the earth's underground makeup, had been significantly improved. All components used in the process were highly refined and delivered exceptional performance. The advent of computers allowed breakthrough innovations.

By conducting several seismic studies of a selected site, computer programs now allowed engineers and geologists to build a three-dimensional image of subterranean formations. Called 3-D seismology, the process uses the reflection of sound waves to produce computer-driven pictures that show reservoirs hidden by complex geological faulting. This ability to see in all perspectives was a totally new exploration tool and resulted in a quantum leap toward doing away with dry holes.

The 3-D imaging technique swept through the oil industry and was quickly adopted. Its effect was enormous. In 1967, on average, 10 wells were drilled to get a single producer. By 1998 that ratio became drill four, get one keeper.

In many ways, 3-D literally put the Gulf of Mexico on the

map, no pun intended. It was a map which showed the entire subsea geology for more miles from shore than anyone had imagined would be productive. In a short time, the Gulf bottom became the most heavily investigated underwater terrain in the world. Today more is known about that acreage than many places on dry land.

The revolutionary 3-D process also had another beneficial effect. Horizontal drilling, or the art of boring straight into the earth, then directing the bit so the hole is steered in any desired direction—left, right, up, or down—was not an uncommon practice. With 3-D, geologists could pinpoint where to place the initial vertical hole and how deep to drill. By locating the exact spot, horizontal holes could then be drilled outward to tap other identified pockets of oil and gas. This process produces the field more thoroughly yet costs less than drilling a number of wells straight down at different places.

In 1989 the Gulf of Mexico provided 15 percent of all oil recovered in the U.S. By 1998 that figure was over 20 percent and growing—thanks to 3-D, which also was making it imperative that drilling techniques keep pace to develop reserves in the deepest of waters.

In 1993 Gulf activity continued to be feverish. During the fourth quarter of that year, Global Marine's worldwide rig utilization was 94 percent. The industry average was 84 percent. Global's higher figure stemmed not only from the quality of its rigs but also from its strong presence in the Gulf.

At the close of 1993 the overall drilling picture was somewhat perturbing. Predictions for 1994 were of an uneven market with sustaining gas well drilling in the North Sea as well as the Gulf of Mexico. Even so, rig activity had been trending upward while rig supply was falling. Another factor which might bring a better market lay in the fact that, since the '86 crash in oil prices, there had been too little exploration outside the Persian Gulf area. Therefore, dependence on oil from the Persian Gulf had increased. Which brought about the possibility that demand for rigs might accelerate.

On a positive note, Global Marine's turnkey drilling program was a success, with income of $10 million on $60 million in revenues. The company had drilled 18 wells, including the very first under a turnkey contract in the North Sea. This method of payment for drilling services clearly had appeal for a segment of the market.

Overall, however, the company reported a small loss for 1993.

During 1994, to fill a forecasting need, Global Marine management created a new economic model which shows the strength of the offshore drilling industry. Called SCORE, for Summary of Current Offshore Rig Economics, this monthly tally reflects current day rates as a percentage of the estimated day rate required to justify new rig construction. Developed by Global Marine vice-president, David A. Herasimchuk, SCORE is a widely respected indicator which reports on a regional as well as rig-type basis.

As 1994 progressed, oil prices began to strengthen. When the year ended, demand for rigs internationally was definitely on the upswing. The Gulf of Mexico also remained active despite lowering natural gas pricing.

At the end of the year, Global Marine earned a profit and built its fleet to 28 rigs. The roster included 24 premium jackups, two third-generation semisubmersibles, a deepwater drillship, and the Concrete Island Drilling System for use in the Arctic.

This gave Global the largest jackup fleet in the world. Day rates for its Gulf of Mexico rigs capable of drilling in 300 feet of water were most advantageous, and in the fourth quarter the 26 available rigs had a 90-percent utilization rate.

All in all, 1994 proved to be productive and satisfactory. For 1995 prospects for the entire industry appeared even better.

During the second quarter, Global Marine purchased the Concrete Island Drilling System (CIDS) on very favorable terms. The company also acquired the rig mounted on the CIDS.

Financially 1995 proved to be the strongest since 1983. The firm made a highly satisfactory profit and management was optimistic. Projections indicated that the market for contract drilling was on a decided upward trend, and the outlook for turnkey business was also promising. With a 96-percent utilization figure, Global Marine sold one rig and ended the year with a fleet of 27 active units going into 1996.

In many ways, the most significant event of 1996 occurred in the last quarter. After months of negotiation, Global Marine announced the return of the *Glomar Explorer* to its fleet. The fabled ship, designed by Global Marine engineers, had earned a place in history. After leasing it from the U.S. Navy for 30 years, Global began converting the heavy-lift vessel to a drillship. Chevron and Texaco had entered into a five-year, $260-million contract for her use and intended to drill in waters deeper than 7,500 feet.

Bringing the *Glomar Explorer* back caused a flurry of media interest. As several articles noted, the *Explorer* was the most famous ship afloat. It was anticipated she would be refitted and on station during the first quarter of 1998.

In a less publicized move, the company also began converting a semisubmersible rig it had acquired, turning a flotel into a drilling vessel. A flotel, as the name suggests, is a floating hotel used to house and care for the needs of engineers and other offshore workers. This one had seen duty in the North Sea and was then moved to Brownsville, Texas, for refitting. Operating under long-term contracts, she would be put to work in the Gulf of Mexico.

In retrospect, 1995 marked the comeback year for the offshore industry as a whole. By the end of 1996, it was apparent that the prolonged down cycle had reversed itself. In fact, 1996, by any measure, was the greatest year Global Marine had ever enjoyed. And 1997 was to be even better.

The company found itself with the best people, the best drill fleet, and a financial strength greater than at any time in the firm's past. Through a rough 10 to 15 years, Global Marine personnel had stayed tough and held their own. Now their perseverance was being rewarded. The company was on course, focused, and well positioned to meet the increasing demand in every market it served.

By the third quarter of 1997, the worldwide utilization

rate of mobile offshore drilling units exceeded 98 percent. Day rates were improving and long-term commitments were common. There was little prospect of a downturn through the end of the century. New marine rigs were being added at a rate of about eight per year, which, because of retirements, meant the total fleet in all waters would remain the same or actually shrink slightly.

Adding to the market strength, a new demand for exploration was being felt throughout the industry. The improved seismic studies made drilling at greater water depths a much more commercially viable venture. Plans were in place to sink wells on the bottom under more than 7,000 feet of ocean. And serious consideration was being given to "ultra deep" sites 10,000 or more feet under the waves.

To position itself for this next challenge, Global Marine began laying out two additional drillships, the *Glomar C.R. Luigs*, named for the president, CEO, and chairman of the board, and the *Glomar Irish Sea I*. Larger than even the *Glomar Explorer*, these new vessels will attain a greater measure of stability through the use of hulls long enough to ride on top of several waves at a time. The added size of each ship, along with all that Global Marine has learned about marine drilling and the very latest equipment, will eventually allow for operations at depths up to 12,000 feet! Dynamically maintaining on-station position, these vessels will have the ability to drill while simultaneously making up casing strings or bottom hole assemblies.

In another nostalgic first, the *Glomar Adriatic IV* arrived in the Santa Barbara Channel on July 28, 1997. The immense unit had been lifted from the water off Gabon in Africa and placed on board a heavy-duty transport, the *Swift*. Carried piggyback for 15,000 miles, the *Adriatic IV* arrived in California and was unloaded. Her task was to work for a combination of six oil companies involved in a program called SWARS, Subsea Well Abandonment and Rig Sharing.

There, off the coast of California, nearly 40 years before, the group which was to become Global Marine perfected the techniques required for marine drilling. The *Adriatic IV* brought the company back to its roots. This was the first time since the 1980s that an offshore rig worked in that area. And its mission was designed to protect the environment by ensuring that closed wells did not become ocean polluters.

Later in the year, on December 9, the *Glomar Celtic Sea* was christened in the AMFELS yard at Brownsville. She entered the fleet and, shortly after, moved to her first job in the Gulf.

At the conclusion of 1997, the Global Marine fleet utilization rate was 100 percent. And the year had been more than profitable for a company which was having to push itself to keep up with demand for its services.

Going into 1998, day rates remained stable or even exceeded those of the previous year. The trend toward lower oil prices began to stabilize at levels which continued to exert only marginal effects on offshore activity.

To meet customer needs, the Global Marine board of directors approved expending $800 million to expand the fleet, with emphasis on deepwater capabilities.

In January 1998, a deal was made to purchase the *Stena Forth*, an enhanced semisubmersible rig designed for use in extreme climatic conditions. Renamed the *Glomar Arctic IV*, she spent a month undergoing refurbishments to be brought up to Global Marine standards before completing her first assignment.

Then in March, Global Marine placed orders for the two drillships which had been in the design stage. These would be the first vessels to enter the fleet, since the *Robert F. Bauer* in 1984, which were built from the hull up expressly for drilling. Called the Glomar 456 class, they will come into service in the fourth quarter of 1999 and the first quarter of 2000. BHP and Exxon have executed long-term contracts for the vessels.

According to announced company plans, Russ Luigs, who had taken the baton of leadership originally held by Bob Bauer, Global Marine's first chairman of the board, was scheduled to retire in May of 1998. Sadly, that event had to be delayed. John G. (Jack) Ryan, the individual designated to assume command of the company, fell ill and underwent a serious operation to remove a malignant brain tumor. Jack had experienced two previous episodes with cancer. Understandably, he elected not to assume the CEO duties.

Fortunately, the quest for a new chief was brief. Robert E. (Bob) Rose began his career in the Gulf Coast office of Global Marine in March of 1964 and enjoyed great success with the company. When the *Glomar II* was sent to Senegal, the ship's manager, in a surprise move, resigned. On short notice, Bob assumed the job and took responsibility for the vessel. After serving in a number of important operations positions, Bob left Global Marine during the 1970s. He became one of the most respected executives in the marine industry and served as president and CEO of Diamond Offshore.

His leaving that company coincided with the unfortunate vacancy at Global Marine. Accepting the challenge, Bob Rose "came home" and assumed the position of president and CEO. Russ Luigs remained as chairman through a period of transition while Bob gathered the reins and became integrated with company operations.

Coincidentally, A.J. Field, Global Marine's first president, had years earlier discussed the possibility of Bob Rose becoming president of Global Marine Drilling. Bob had liked the idea but suggested his office should be in Houston. At the time, the move was rejected although, two years later, Global Marine relocated the company headquarters to that city.

Chapter 14

The Deep Frontier — Looking Forward

Predictions of running out of oil have been made since the late 1800s. Pundits, usually backed by special-interest groups with an axe to grind, periodically appear in the popular press to forecast the end of our petroleum resources. The fact that it has not happened does not seem to discourage this particular kind of prognostication.

To get a more reasoned view of what the future holds in terms of meeting our hydrocarbon needs and to see where offshore drilling is heading, it is necessary to grasp some simple fundamentals.

As those in the oil industry well understand, establishing a figure for known petroleum reserves without having a per-barrel price as part of the equation is fruitless. The production of oil and gas is a business. And no business can survive by operating at a loss. Which is what an oil company would do if it paid more per barrel of oil or cubic foot of gas than could be recovered when the commodity was sold.

Untold amounts of hydrocarbon are available to us through the processing of coal, and even more is there in the form of tar sands. Extracting the hydrocarbons from these sources is very expensive. The cost is so great, in fact, that this form of refining has not been performed on a large commercial scale because the resulting product would have to be priced above oil on the open market. If the price of oil were high enough, or if a new technology allowed cheaper conversion, these resources would be readily accessible.

The same is true for petroleum. There are unfound, untapped fields. Some will be expensive to produce, so will remain unused until the economics are right.

These examples make it clear that, for a given price per barrel, there is a given volume of known hydrocarbon reserves. Raising the price per barrel of oil increases the reserves. Lowering the price per barrel of oil reduces the reserves because it is not commercially feasible to recover them for that amount of money.

Since the onset of supply and price instability, which began with the OPEC embargo in the 1970s, exploration for new sources of oil has caused a major shift in production. While U.S. output has slipped a bit, to about 7 million barrels a day, the balance of world production outside the OPEC nations rose from nearly 20 million barrels to almost 35 million barrels per day. OPEC production has varied from around 25 million barrels to 30 million per day.

A significant portion of the production from non-OPEC nations has come about through offshore exploration. And there is every reason to believe marine drilling will play a larger role in finding and producing petroleum in years to come. Because, even with all the offshore wells that have been drilled in the past, the seafloor, outside of a few areas, remains a vast uncharted frontier.

Demand for oil and gas is increasing. And while we are not running out of reserves, worldwide excess oil production capacity, as a percent of total actual production, is shrinking. In 1983 it would have been possible to produce 37 percent more oil than was being taken from the earth. In 1998 that figure dropped to only 7 percent. Those numbers reflect both use of more oil and the fact that exploration has not been proceeding at a sufficiently rapid pace.

Excluding the former Soviet Union, hydrocarbon consumption increased 35 percent in the 10 years from 1987 through 1997. If this trend continues, and in all probability it will, projections indicate a growth in consumption of about 30 percent in the period 1997 through 2007. This means an increase in production of close to 30 million barrels per day by 2007. On top of that, we will have to recover another 25 million barrels daily to replace production lost through depletion of existing wells in the next 10 years. This adds up to 55 million barrels per day. And since excess production is currently only about 7 percent, most of that 55 million barrels will have to come from new sources.

Will new sources be there?

At what price?

For maximum exploration and production activity, a price high enough to reward the effort and low enough to stimulate consumption is ideal. In today's marketplace, that price is well below $30 per barrel. In fact, oil prices of $30 and above would be a signal for those holding shares in oil field drilling and supply companies to place a careful watch on their investments.

In any case, at the proper price, new sources can be found and developed to meet the coming shortfall. And many of these new sources will be in ultra deep water, which is currently considered depths more than 10,000 feet. Emerging areas will include the Gulf of Mexico, farther from the U.S. and Mexican shores; the coast of South America off Brazil; and offshore West Africa.

Ultra deep water means greater costs for a new well. Which in turn indicate that each barrel of new oil will be more expensive at the wellhead. However, if the price goes too high, demand will decline. And if demand falls, the cycle will begin again.

The wild cards in this scenario are the advancements in technology.

The story of marine drilling provides a clear demonstration of technical achievement. Inventive minds working with oil company managers who are willing to take a chance in order to find a better, less expensive way thrive in this industry.

While no one will be able to overcome the fact that drilling in deeper water will be more costly than drilling at shallower depths, improved exploration techniques may help moderate that difference.

As noted, 30 years ago, having one successful well out of 10 tries when drilling in unproven territory was expected. Today with the advent of 3-D seismology, the ratio is one hit out of four attempts. The end result is that a new technology significantly reduced the number of dry holes and therefore lowered the costs required to find new oil and gas.

A group of 27 oil companies has banded together to develop what is called 3-D vector wavefield seismic. Working as a team, in a fashion similar to the CUSS operation of the 1950s, the member firms support this effort financially and with equipment as well as personnel. This promising advance reportedly gives an even better 3-D view of the prospect and enhanced recovery from the field.

In another effort, a system known as 4-D seismic adds a time factor to 3-D pictures, giving a view of the migration of oil and gas deposits in a field. This too aids in greater recovery.

While neither of these processes is perhaps as revolutionary as 3-D seismology when it was introduced, both are meaningful enhancements.

Still to come are basic improvements in the types of signals used to generate seismic studies and the sensitivity of receivers that record those signals. Great cost savings would be realized by the ability to make a complete survey in a single pass over an area.

Another huge economy, along with the capability of drilling effectively at depths of 15,000 or more feet, will be found in eliminating the marine riser. The riser is now used to connect a well to its surface platform. Work is in progress to perform tasks, such as drilling fluid circulation and disposal of the hole cuttings, through seabed-mounted equipment. These advances will shorten the time required to drill to a given depth.

Improved drill bits, which last longer and are capable of "reading" the formation ahead as hole is being made, can also speed up the process and cut costs. Drill pipe on a spool which comes off in a continuous length is another innovation now under development.

For deepest waters, the drillship may well be followed by another type of platform resembling a spar. The spar is similar in many ways to a concept considered by the original CUSS group decades earlier, called a "lily pad." The work area, drilling floor, derrick, crew quarters, and other accommodations are mounted above a single triangulated hull in the shape of an open spar which extends far down into the water, resembling a huge stalk. Chambers in the spar can be flooded, like a semisubmersible, to lend additional stability in rough weather. The spar is held in place by tensioned tendons connected straight to the bottom or by a wide-based mooring system.

Research is currently underway in every phase of the offshore industry. No part of the process of exploration and production is being overlooked. Even environmental issues, which are sure to arise as more of the activities are carried out beneath the water, are being considered. Molecular chemists are progressing with the conversion of spilled oil, exhaust gas, effluents, and other emissions into nontoxic, neutral substances.

Advances will, if the past is any indication, continue to improve performance, extend the deepwater frontier, and allow greater, more complete depletion of every newly discovered field.

If someone in the offshore industry today were to do a Rip Van Winkle and awaken 20 years in the future, he or she would see a great many changes. One constant, though, would be the people.

Marine drilling has been made possible because of the contributions of thousands of dedicated people. The men who toil in arduous conditions with little protection from the elements, and those who support their efforts, all strive to accomplish a single task. They seek to bring us energy from the seabed. For those who work in the office and those who venture into the waters, hunting and producing petroleum is more than just a way to make a living. It is a lifestyle in itself, which attracts a special breed. The coming advanced technology is important. But maintaining the cadre of skilled, enthusiastic people who will invent, build, and use that technology is vital.

To paraphrase Winston Churchill, seldom have so many owed so much to so few.

As a final salute to those intrepid folks, consider this true story. The details are offered as they appeared in the news media and the actions speak for themselves.

It was 12:30 on a dark, hot night on the Gulf of Mexico. Some 100 miles south and east of Cameron, Louisiana, 42 workers on a platform faced a deadly choice. Their well had blown out. Escaping gas had turned the sea into a frothy cauldron, and the rig was engulfed with the explosive vapor. Staying on the shuddering, banging rig amid the debris meant risking death by fire and explosion.

Escaping into the water required climbing or lowering themselves nearly 100 feet from the main deck, then facing terrors of the unknown. Assuming survival after splashing into the foaming sea, would they be able to get far enough away from the rig before it was engulfed in a fireball?

The chance of aid from a rescue ship seemed remote. One vessel they had managed to contact refused to approach the potential bomb, and with good reason. Aside from the threat of an explosion, there was definite risk of sinking from another cause. When gas is forced into the water under great pressure, millions of tiny bubbles are formed. This infusion changes the density of the water, and in many cases, boats cannot remain afloat.

A second vessel, a few miles from the troubled platform, also heard the distress call. Knowing the rig was "kicking" or shaking as pockets of gas slammed upward from the earth, the crew responded.

The captain and his four hands turned from their course and headed straight for the endangered men. The 175-foot supply ship smashed through waves 10 feet high into 20-knot winds and came alongside one of the support legs. Shouting to be heard, and using the radio, they announced their arrival, pleading with those on the threatened rig not to jump into the water. In the dark and confusion, finding a person amid the waves would be difficult if not impossible.

The experienced captain slipped his ship into position and held it there against the rough sea. Despite the gas and water mixture, they were still afloat, which was some relief. The crew began to help those scrambling down ropes and ladders. A few of the rig hands struggled through part of the distance separating them from relative safety, then fell or jumped onto the pitching deck. Bodies flying from above added to the chaos and danger for those below.

When the last man came aboard, the captain, sweating, added power and the ship pulled away. Just as he was relaxing a little, there was a startled shout. The men had done a roll call and they were one short. Someone was still aboard the doomed rig.

Without hesitating, the ship turned back into danger. A single spark could ignite an inferno. The boat crewmen, working with disciplined control, trained a spotlight on the rig floor, scanning from side to side trying to locate the missing worker. Time passed before they found him. With tension building, he was helped aboard and they turned to make their second escape.

Less than 30 minutes later, as the crew handed out what blankets and spare clothing they had, a brilliant fireball blazed into the night sky behind them. Then across the water came the concussion and noise of a powerful explosion. The resulting fire melted the rig to the water line.

Thanks to the boat crew's daring and courage, no lives were lost. Disaster had been averted.

Thankfully, stories like this are not common in the offshore industry. This one, though, clearly captures the spirit and mettle of those who respond to the deep challenge.

Appendix

How to Drill an Oil Well in the Middle of the Ocean: A Quick Primer

Drilling an oil well at sea is a very different matter from doing the same job on land. Marine drilling is more complex, uses a great deal of special equipment, and costs are considerably higher. However, the basic principle of making the hole is much the same. It's a straightforward process.

In the simplest terms, a drill bit is screwed onto the end of a heavy length of pipe called a drill collar. One or more collars may be used depending upon how much weight is to be placed on the bit. The collar is then lowered with the bit toward the spot where the hole is to be made. When the full length of the collar is extended, it is held in place while a section of drill pipe is screwed into it. This process continues until the bit makes contact with the seabed. Then the drill pipe is rotated, which turns the collar, turning the bit, which makes the hole.

A length of drill pipe is usually 30 feet. So when the bit has bored deep enough, another drill pipe must be screwed or "made" onto the first. This is called "making a joint." The two drill pipes, the collar, and the bit are then a "drill string."

When the drill bit penetrates further, a third length of drill pipe is made onto the second, and when necessary, a fourth made on to the third, and so on. Want to drill down 6,000 feet? Repeat the process 200 times. Need to save time and money drilling down 6,000 feet? Make up (screw together) three sections of drill pipe and treat it as a single 90-foot length, or "stand." A stand means handling fewer sections of pipe and having to make or "break" fewer joints when going into or coming out of the hole, which is a process called "tripping."

The drill bit does not auger into the seabed. It grinds its way deeper and deeper. Modern bits are made of advanced metals and have several sets of tough serrated teeth mounted on conical rollers.

As the bit is turned, the teeth rotate, and each one takes a little bite out of the earth. Even the toughest rock cannot resist the gouging, gnawing action.

Turning the bit by twisting the drill string requires a great deal of power and some form of rotary system.

On the *Glomar Explorer*, a Global Marine drillship designed to work in very deep waters, all power is provided by a diesel-electric system, much like that used in a train locomotive. Several huge diesel engines run economically and reliably at a fixed speed. Each is connected to a generator, which produces electricity. Electricity powers motors that turn the drill string, drive the ship, and run other necessary machinery on board, including air conditioning.

In the past, the drill string was turned by a "rotary table" mounted on the rig floor. Today on the *Glomar Explorer*, the rotary table is still there but is only used to guide and handle the drill string. In the interests of safety, speed, and expense, an electric motor mounted high in the derrick twists the bit.

Using this top drive system also allows the entire process of making up or breaking out the drill string to be automated. A top drive system and an "iron roughneck," which is a pipe-handling robot built to operate on the rig floor, eliminate having people exposed to injury in that dangerous area. Need for this safety measure is especially critical when working offshore on an unanchored drillship. Storing or "racking" the lengths of drill pipe on their sides, as opposed to hanging them on end, is another security measure. Global Marine pioneered this innovation.

As the rotating bit chews away, chips of rock called "cuttings" are created. If the cuttings stayed in the hole, they would clog the hole and keep the bit from reaching the bottom where it can drill deeper. So the cuttings must be removed.

To do this, a slurry of water, chemicals, clays, and other minerals is used. Called "drilling fluid," or more often "mud," this thick liquid is pumped under pressure down the inside of the hollow drill pipe. It emerges through special outlets on the bit as powerful jets that clean the bit teeth. The expelled mud, forced by an unending pressurized supply pouring through the drill string, moves upward in the space between the outside of the drill string and the wall of the hole, carrying the cuttings with it.

Free of cuttings, the bit makes contact with fresh rock and continues gouging, burrowing deeper and deeper.

The contaminated mud, coming back up out of the hole, reaches a mud-return line, which is a pipe that vents it from the hole. Mud flows through this pipe onto a "shale shaker." The shaker is a machine with inclined vibrating screens or sieves. Sieves separate mud from cuttings, allowing the mud to reenter the mud-storage tank called the "mud pit." Cuttings and other material strained from the returning mud are discarded. From the mud pit, a suction line constantly sucks the cleaned mud to the mud pump, where it is then forced back down the hole to repeat the process.

To handle the massive weights involved when raising or lowering the drill string and to gain enough height to have working room, a hoisting system is needed. Operated by steel wire ropes and driven by an electric motor is a winch or hoist, called the "drawworks." The wire rope is wound through a set of pulleys to multiply the force needed to lift or lower many tons. To support the lifting-lowering process, a tall derrick, or open, trussed steel tower, is required. The drill string is literally hung by cables from the top of the derrick and is actually suspended in the hole.

To accomplish these tasks at sea presents some intriguing challenges. The following is a simplified outline of the process used aboard the *Glomar Explorer*, which can work very deep waters.

One of the first differences in marine drilling is that there are no landmarks. So satellite navigation and other electronic aids are used to place the ship over the exact spot where the well is to be sunk. Since the water may be 5,000 to 10,000 or more feet deep, anchoring the ship to hold it on site is impossible.

Once on station, special sonar units are lowered to the seafloor. Their signals are picked up by receivers on the *Glomar Explorer's* hull, and information they provide is fed into computers. The computers use this data to detect even the slightest deviation in position and trigger props as well as giant thrusters into action. A thruster is a huge, electrically powered propeller mounted below the ship which provides side-to-side mobility for the vessel. Thrusters quickly nudge the ship back into exact alignment and then hold it there.

After positioning is assured, drilling can begin.

The *Glomar Explorer*, like other Global Marine ships, is equipped with a special opening called a "moon pool." Global Marine invented this concept, which is an aperture through the vessel from the drilling floor directly into the water. When deployed, the drill string hangs straight down from the derrick through this opening. The moon pool, despite its romantic name, is a busy work area.

The next step is to set a temporary guide base on the seabed in exactly the desired spot. Placement is done by attaching the heavy steel base to the end of a drill pipe and lowering it into position. The temporary guide base has guidelines running upward to the ship.

Next, a pair of bits are made onto drill collars and then attached to the drill string. The first bit makes a hole less than 18 inches across. The second bit, called a "hole opener," widens the hole so that steel pipe, called "casing," can be slipped into the hole to keep it open. The drill string is lowered through a "guide frame," which runs on the guidelines that stretch from the temporary guide base to the ship. The cables "guide" the drill bit into an opening on top of the guide base. Once in place, the drill string starts turning and a hole is made.

After about a hundred feet or so, the bits are brought back to the surface and the casing is lowered into the hole, using the same guidelines. Following successful placement of the casing, the permanent guide structure is let down and landed on its temporary cousin, which stays in place and acts as a base. The hollow drill pipe is then lowered into the hole and cement is pumped down through the pipe to harden and lock the entire structure in place.

The process of drilling, then positioning and cementing casing is continued until there is a metal-lined hole through the softer sediments down to firmer material. This technique prevents the hole from being closed or filled by the sides caving in. The "conductor casing" as it is called, along with its concrete reinforcement, is the foundation for the well.

For safe drilling, on land or sea, a massive set of high-pressure valves known as a "blowout preventer" is mandatory. If the drill bit penetrates an area of very high pressure, sending oil, gas, or both, back up the hole with tremendous force, the result is called a "blowout." Blowouts can cause great damage. Fire is a real hazard if the flammable mixture reaches open air out of control. And a fire at sea is an extremely serious matter. The blowout preventer, or BOP stack, is a series of valve-like devices stacked one on top of the other. In case of a high-pressure situation, the BOP stack seals off the hole and the pipe, preventing a blowout.

Once the permanent guide structure is set and cemented into place, the BOP stack is lowered by the drill pipe and secured to the permanent guide structure. Hydraulically and electrically connected to the ship, it is the safety valve in case of trouble.

At this point, if the drill bit entered the hole and drilling started in earnest, the mud that was pumped in would flow upward, out of the hole into the water and never be recovered for reuse. So a pipeline is laid from the BOP stack, which sits on the well, upward into the moon pool of the ship. Working in 10,000 feet of water, the pipeline, called the "marine riser system," is 10,000 feet long. It is lowered, one length of riser pipe at a time, until it is connected to the BOP stack. Obviously, a two-mile string of pipe is exceptionally heavy, so it is suspended from the ship by a series of cables to hold it in tension and prevent it from bending or collapsing under its own weight. In very deep water, special flotation devices are used to reduce the tension on the riser line.

In a normal operation, all the above must be accomplished before any drilling begins. The process involves a lot of work, and no shortcuts are available.

To assist in this effort, various undersea devices are employed. In water less than 1,000 feet deep, divers can work in special hardsuits or enclosed habitats. And for tasks at greater depths, there are unmanned submersibles which are controlled from the surface. These Remote Operated Vehicles (ROV) have arms which can be fitted with tools to perform various jobs. Vision in the depths is also helpful. Global Marine was the first to utilize undersea TV cameras for drilling. These can be used to inspect the marine riser, check installation of the blowout preventer stack, and help position objects on the seafloor without the aid of guide wires.

In any case, when all is complete, one other item is needed to make sea drilling possible.

Imagine a drill bit on a string 12,000 feet long in 10,000 feet of water. It hangs down straight from the derrick through the moon pool, into the riser line, through the BOP stack, and then down 2,000 feet into the hole until the bit is turning against the bottom.

On the surface, a large swell or wave comes along and lifts the ship 12 feet, then the vessel settles into the trough. Down in the hole, the drill bit would be lifted up 12 feet, following the motion of the ship, then slammed down against the bottom of the hole with the force of a train wreck. Clearly, this is not good for the ship, the marine riser line, or the drill string and would result in catastrophic damage.

To prevent massive movement, a hydraulically operated device, the "heave compensator," makes up for the up-and-down motion of the ship, keeping the connection between the drill string and the derrick at a constant distance from the seafloor. Global Marine's early work in this area alone was one of the factors that made deepwater drilling from a floating ship (a "floater") possible.

The above description is greatly simplified. It requires days or even weeks of preparation before a bit can be turned to make a well in deep water. And a great many people are involved.

Key players are the toolpushers, who lead the drilling crews; the drillers, who operate the controls and make the hole; roughnecks, who assist the driller; mechanics, who keep equipment working properly; mud specialists, electricians, crane operators, cooks, galley help, and roustabouts for general work; the nautical crew, the geologists, an office staff back on land, helicopter pilots, supply boat workers, and the catering-housekeeping staff, to name only a few.

These men and women have skills which have been honed to a level of excellence. Their endurance has been tried and their courage tested. The work they perform is invaluable to our society and civilization.

A Very Warm Thanks

A very warm thanks to the many fine people who made this book possible.

I especially appreciate the time Bob Bauer, A.J. Field, Curtis Crooke, Hal Stratton, Taylor Hancock, Russ Luigs, and Bob Rose were kind enough to spend with me in person and over the telephone. They lived this story and know far more about it than I.

I would also like to thank Dave Herasimchuk for his very special input.

And Sherman Wetmore certainly needs to be recognized for the excellent historical retrospective he created several years ago.

The fine writers and editors at *Offshore* magazine, and particularly Marshall DeLuca and Leonard Le Blanc, are deserving of grateful recognition. They graciously provided access to their library facilities during the research phase of this project.

It is also appropriate to recognize Connie Abbs, Audrey Darden, Joanne Ozment, and Janice Burge for their prompt and efficient aid.

Dennis Batham, Ray Perry, George Helland, and John Duke are four more who volunteered information.

And finally, a tip of the hat to Ocean Star Museum, a wonderfully interesting facility dedicated to marine oil and gas activities. The Ocean Star is housed in an offshore drilling platform at Galveston, Texas, and is open to the public.

Index

The page reference P *indicates the section of photographs.*

D

E

H

Halliburton, 171
Hancock, Taylor, 110
Hannah, A. Douglas, 103
Harbor Boat Yard, 40
Harrell No. 7 disaster, 25-26
Havenstrite interests, 120
Hayward, John T., 74-78
heave compensator, 89, 202
helicopters, 110-111
Herasimchuk, David A., 184
Hess, Harry, 86
Heywood No. 1, 33
High Island class rigs, 160
High Island exploration, 163,
    179-180
Hill, Ken, 133
Hillman & Sons Company, 103
Hillman, Henry L., 103
HMS Challenger, 125
hoisting system, 199
hole opener, 200
Honeywell Corporation, 97
Hong Kong financing, 160
Hoover, Herbert, 47
horizontal drilling techniques, 183
Houston, and oil bust, 174
Hughes Air West, 143
Hughes Glomar Explorer, P,
    see Glomar Explorer
Hughes, Howard, 139-141, 143-146
Hughes Marine Barge, HMB-1, 140,
    141, 142
Hughes Tool Company, 120, 140
Humble Oil & Refining Company,
    5, 10, 66, 108
Hunt brothers, 119
Hurricane Andrew, 180
Hurricane Betsy, 10
Hurricane Camille, 1-11
Hurricane Hunters, 4-5

hurricane threat, 70
Hutton field, 82

I

Ickes, Harold, 47-48
illumination, methods of, 14, 15, 17,
    18-19, 22-23, 24
Incentive Drilling Program (IDP),
    178-179, 180
industrial revolution, 15
inland-barge rigs, 82, 83
Institute of Marine Science, 124
Interior and Insular Affairs
    Committee, 53
International Union of Geodesy and
    Geophysics, 87
iron roughneck, 198

J

jackup drilling rigs, 82, 83, 114,
    158, 160, 163, 164, 178,
    181-182, 184
Jade lease, 66
Jennifer Project, 139-146
jet coring, 41
J.H. Whitney & Company, 103
Johnson, Lyndon Baines, 92-93, 99
JOIDES (Joint Oceanographic
    Institution for Deep Earth
    Sampling), 124-132
Julie M., 44

K

K-129, 135-138, 145-146
    see also Jennifer Project
Kennedy, John F., 53, 92
Kermac-16, 73
kerosene, 17, 19, 23
Kerr-McGee Oil Industries, 69, 76,
    77

offshore coastal oil, *see* tidelands controversy
offshore diving, 45, 60-61, 81, 82, 142, 156, 170, 202
offshore drilling
    beginnings of, 24, 37
    market for, 101, 113-114, 150, 151, 157, 158, 160, 161, 164, 183, 184, 186, 190
    objections to, 62-63
offshore living quarters, *see* living quarters on rigs
offshore technology, beginnings of, 43-45
offshore windmill farm, 162
oil
    market for, 149, 153, 154, 157, 168, 175, 190
    see also production of oil
Oil City, 26-27
oil embargoes, 150, 152-154, 157, 190
oil industry downturn, 168-169, 174
oil wells
    how to drill, 197-203
    inception of, 20-22
omnidirectional waterjet, 132
OPEC (Organization of Petroleum Exporting Countries), 152-154, 172-173, 190
open-truss configuration of rig legs, 83
Ortega Rancho, 24
OSHA (Occupational Safety and Health Administration), 57
OTC (Offshore Technology Conference), 164

P

Pacesetter class rigs, 161
Pan American International, 108
paraffin, 16

Pennsylvania Rock Oil Company, 18-21
People's Republic of China exploration, 169-171
Persian Gulf oil production, 183
petrochemical belt, 69
petroleum
    beginning of industry, 13-31
    distilling of, 17, 18
    as fuel, 13, 22, 23, 28
    impact of, 13-14
    international aspects of business, 28, 149
    medicinal claims for, 16, 18, 19
    refining of, 16, 17
    volatility of, 17, 18, 25-26, 201
    see also production of oil
Phillips Petroleum Company, 69, 71, 87, 109
phosphate deposits, recovery of, 118
plate tectonics, 88, 124, 129-130
platform-tender systems, 72-73
platforms
    designs of, 71-80, 192
    on pilings, 34, 37, 38, 71
pneumatic pipe-handling system, 59
Polaris missile test facility, 68
Polesti oil fields, 39
politics, and Project Mohole, 92-93, 95
Poly Glomar Driller, 154
posted-barge rigs, 76, 83
posted-submersible barges, 77-78
price
    of natural gas, 179, 180, 184
    of oil, 152-154, 157, 168, 172-173, 184, 187, 189-191
Producers No. 2 (Caddo Lake), 25
production of oil
    in early days, 22
    and pricing, 152-154, 190-191
    in U.S., 55, 149, 183, 190
    in Venezuela, 28, 29

worldwide, 160
see also oil embargoes
Project Mohole, 85-100, 123, 127
Pure Oil Company, 37, 69, 108
purpose-built vessels, 117
Pyron, W.B., 27, 28

Q

el-Qaddafi, Muammar, 150
Quintana Overseas Inc., 152

R

racking of drill pipe, 58-59, 198
recirculating head, 44
Red Sea minerals, 131
reentering the hole, 88, 132
reflective seismography, 39
Remote Operated Vehicle (ROV),
    61, 202
Republic of Texas, 49
reserves of oil, 37, 183, 189-190
retractable leg platform, 79-80
Richardson, Sid, 51-52, 54
Richfield Oil Company, 66
rig downtime, 104
Rig 44, 77
Rig 54, 77
Rigger I, 62
Rincon, 66
riser pipe, 60
    see also marine risers
Rita Candies, 5
Robert F. Bauer, 187
rock oil, 14, 18, 20, 23-24, 31
Roosevelt, Franklin D., 47, 48
Rose, Robert E. (Bob), *P,* 114, 188
rotary tables, 43, 198
roughnecks, 203
Rouse, Henry R., 22
Rumanian oil fields, 39
Ryan, John G. (Jack), 187-188

S

Sadat, Anwar, 152
salt domes, 33-34, 70, 129
salt wells, 15-16, 19, 20
Santa Barbara Channel oil spill,
    150, 156
Santa Fe Company, 57, 64
satellite navigation and communica-
    tions system, 127, 128,
    199-200
Saudi Arabia, 172
Schlumberger, 171
Schmidt, Benno C., 103
SCORE (Summary of Current
    Offshore Rig Economics),
    184
Scorpion, 80
Scripps Institution of Oceanography,
    124-129
Sea Spider, 136
Seaboard Oil Company, 75
seafloor spreading, 124
seaweed as alternative energy, 162
SEC (Securities and Exchange
    Commission), 143-145, 163
seismic surveys, 182-183, 186, 192
seismograph, 39
self-powered ships, 73, 104-105,
    107, 109
self-propelled semisubmersible
    platforms, 97
semis, *see* semisubmersible rigs
semisubmersible rigs, 82, 83-84,
    97-98, 154, 157, 158, 160,
    161, 164, 184, 185, 187
Seneca Oil Company, 21
shale shaker, 199
shallow-water production systems,
    165
Shell Oil Company, 28, 37, 40, 66,
    69, 82, 89, 94, 108
    and *CUSS II,* 101-107
    see also CUSS Group

216